MATLAB Engineer's Toolkit

Volume 2

Engineering Mathematics, Data Analysis, and Economics

MATLAB® Functions for Linear Algebra, Numerical Methods, Statistics, Experimental Data, Engineering Economics, and General Engineering Utilities

Peter I. Kattan

Petra Books

www.PetraBooks.com

This book is intended for educational and informational purposes only. The author and publisher make no guarantees regarding financial, business, or personal outcomes resulting from the application of the information presented in this book.

ISBN: 979-8-1827-1924-7

Published by Petra Books

Printed in the United States of America

First Edition

Contents

Preface

Modern engineering requires more than the application of physical formulas. Engineers routinely solve systems of equations, work with vectors and matrices, apply numerical methods, analyze experimental data, quantify uncertainty, fit models to measurements, and evaluate the economic consequences of engineering decisions.

MATLAB is particularly well suited to these tasks. Its capabilities in numerical computation, matrix operations, data analysis, and scientific programming make it an important tool in engineering education and professional practice. Many useful engineering calculations can be implemented as short MATLAB functions that are easy to understand, verify, modify, and reuse.

This book, **MATLAB Engineer's Toolkit, Volume 2: Engineering Mathematics, Data Analysis, and Economics**, presents **59 MATLAB functions** covering engineering mathematics, numerical methods, statistics and experimental data, engineering economics, and general engineering utilities.

The book is organized into four major parts and nine chapters.

Part I — Engineering Mathematics covers algebra, vectors, matrices, and numerical methods. Functions are provided for quadratic equations, systems of linear equations, matrix operations, eigenvalues and eigenvectors, vector calculations, interpolation, numerical differentiation and integration, nonlinear root finding, curve fitting, and linear regression.

Part II — Statistics and Experimental Data addresses the analysis and interpretation of engineering measurements. It includes descriptive statistics, error and uncertainty calculations, confidence intervals, correlation, moving-average smoothing, data normalization, and outlier detection.

Part III — Engineering Economics introduces computational tools for the economic evaluation of engineering decisions. Topics include interest, present and future value, loan payments, annual cost, net present value, payback period, break-even analysis, depreciation, project profit, profit margin, and basic cost estimation.

Part IV — General Engineering Utilities contains broadly useful functions for engineering prefixes, scientific notation, significant figures, percentage change, ratios, efficiency, table interpolation, design and safety margins, engineering constants, material properties, unit references, and batch calculations.

As in Volume 1, every function follows a consistent format. The mathematical or engineering basis of the calculation is presented first, followed by definitions of the variables and units where applicable. The complete MATLAB function is then provided, together with a short explanation and a numerical example.

The functions are intentionally concise. Their purpose is not simply to produce numerical answers but also to demonstrate how mathematical and engineering procedures can be translated into clear MATLAB code. Readers are encouraged to examine and modify the programs rather than treating them as black-box calculations.

The book is intended for engineering students, instructors, researchers, and practicing engineers. Students can use it to strengthen their understanding of engineering mathematics and computational methods while gaining practical MATLAB experience. Instructors may use the functions as teaching examples or exercises. Engineers and researchers can employ individual routines for calculations, data processing, preliminary analyses, economic comparisons, and as components of larger MATLAB programs.

This volume complements **MATLAB Engineer's Toolkit, Volume 1: Fundamental Engineering Calculations**, which contains 64 MATLAB functions covering engineering geometry, unit conversions, temperature conversions, and basic thermal calculations.

The two volumes are designed to stand independently. Together, however, they form a collection of **123 MATLAB functions** spanning a broad range of fundamental engineering computation.

Our goal throughout the series is to keep the relationship between engineering theory, mathematical formulation, MATLAB implementation, and practical application as clear as possible.

We hope that the functions in this volume will serve not only as ready-to-use computational tools but also as foundations from which readers can build their own MATLAB libraries for engineering analysis and problem solving.

Peter I. Kattan

September 2026

Copyright

The MATLAB functions presented in this book are provided for educational, instructional, and general engineering purposes. Although every effort has been made to ensure the accuracy of the material, the authors make no warranty, expressed or implied, regarding the completeness, accuracy, or suitability of the functions for any particular engineering application.

Users are responsible for verifying all calculations and for ensuring compliance with applicable engineering codes, standards, regulations, specifications, and professional requirements. The functions in this book are not intended to replace professional engineering judgment or specialized engineering software where such software is required.

How to Use This Book

This book has been designed as a practical reference rather than a text that must necessarily be read sequentially. Each MATLAB function is presented as a largely self-contained calculation, allowing readers to proceed directly to the function required for a particular problem.

Standard Format

Each MATLAB function follows the same basic structure.

1. M-File Name and Description

The MATLAB filename is presented together with a concise description of the calculation. The function should normally be saved using the specified .m filename.

2. Main Equations

The principal mathematical or engineering relationships used by the function are presented before the MATLAB code. This allows the reader to understand the basis of the calculation before examining its computational implementation.

3. Variables and Units

The variables appearing in the calculation are defined together with their appropriate units. Dimensionless quantities are identified where appropriate.

4. MATLAB Function

The complete MATLAB function is then presented. The programs use meaningful variable names, clear comments, and basic input validation where appropriate. MATLAB code is displayed in a monospaced typeface to distinguish it clearly from the surrounding text.

5. How the Code Works

A short explanation describes the principal steps performed by the function and identifies any important MATLAB operations.

6. Numerical Example

Each function concludes with a numerical example showing representative input values, the MATLAB function call, the calculated output, and the final result.

Saving the Functions

Each function should normally be saved in a separate M-file using the filename given in its section heading. The file should be placed in the MATLAB Current Folder or another folder included on the MATLAB search path.

Readers who use many of the functions may wish to create a dedicated folder for the **MATLAB Engineer's Toolkit** and organize the functions into subfolders corresponding to the chapters or subject areas.

Using the Functions

A function may be used independently for a single calculation or incorporated into a MATLAB script or larger function.

The functions are intended to be computational building blocks. Readers are encouraged to modify them for their own requirements, including processing arrays of values, generating tables, producing plots, performing repeated calculations, or combining several functions into a larger engineering procedure.

Engineering Units

Unless a function is specifically intended for unit conversion, MATLAB does not automatically determine the physical units associated with numerical inputs. The user is responsible for maintaining a consistent system of units.

Units should therefore be checked before entering data and again when interpreting the calculated result.

Verification

A computer-generated result should never be accepted solely because MATLAB produced it without an error message.

Readers should verify important calculations by checking units, expected magnitudes, signs, limiting cases, and known results whenever possible. Engineering judgment remains an essential part of computational engineering.

For safety-critical or regulated applications, the functions should be independently verified and used in conjunction with applicable engineering codes, standards, specifications, and professional requirements.

MATLAB Requirements

The functions in the **MATLAB Engineer's Toolkit** have been designed to require only a basic working installation of MATLAB whenever practical.

Most functions use standard MATLAB capabilities such as scalar arithmetic, vectors, matrices, mathematical operations, conditional statements, logical operations, and user-defined functions. Specialized programming techniques and additional toolboxes have been avoided unless they are genuinely necessary for a particular calculation.

A reasonably recent MATLAB release is recommended. Because most functions rely on fundamental MATLAB syntax, many should also operate with older releases. Readers using an older version should verify that any commands used by a particular function are supported by their installation.

Basic Requirements

To use the functions in this book, the reader should have:

- a computer capable of running MATLAB,
- a working MATLAB installation,

- access to the MATLAB Editor and Command Window,
- permission to create and save .m files, and
- basic familiarity with entering MATLAB commands and calling functions.

No high-performance computer is required for the calculations presented in these volumes.

MATLAB Toolboxes

The majority of the functions are intended to use **base MATLAB**. If a particular function requires functionality associated with an additional MATLAB product or toolbox, this should be identified in the corresponding section.

M-Files

Each MATLAB function is normally stored in an individual M-file. The filename should correspond to the name of the primary function contained in the file.

Using descriptive filenames makes the collection easier to organize and allows individual functions to be located quickly within a personal MATLAB library.

Numerical Precision

MATLAB normally performs numerical calculations using double-precision floating-point arithmetic. This is sufficient for the calculations presented in this series.

The number of digits displayed on the screen may be smaller than the number retained internally by MATLAB. Final engineering results should therefore be reported with an appropriate number of significant figures based on the accuracy of the input data and the requirements of the application.

Internet Connection

An internet connection is not required to perform the calculations once MATLAB is installed and properly licensed. Internet access may, however, be required for software installation, licensing, updates, or access to online documentation.

Organization of the MATLAB Functions

The functions in each volume are organized primarily according to **engineering subject**, rather than according to MATLAB programming commands.

This approach allows readers to search for the engineering calculation they need without first knowing which MATLAB technique is required to perform it.

Each volume is divided into parts and chapters containing related calculations. Within each chapter, individual MATLAB functions are presented as separate, self-contained sections.

Function Naming

Descriptive filenames are used whenever practical. MATLAB filenames do not contain spaces; underscores are therefore used to separate words.

A filename is intended to communicate the purpose of the function as clearly as possible, making the collection easier to navigate from both the book and the MATLAB file system.

Independent Functions

Each function is designed, whenever practical, to operate independently of the other functions.

A reader should therefore be able to select one function from the book and use it without installing or executing the entire collection.

This independence is important because it allows the book to function as a reference library rather than requiring sequential use.

Combining Functions

Although individual functions are independent, they may also be combined.

The output from one function may become the input to another, several functions may be called from a MATLAB script, or individual functions may be incorporated into a larger engineering program.

This modular organization allows simple computational tools to become components of increasingly sophisticated engineering procedures.

Function Indexes

The volume includes indexes designed to make individual calculations easy to locate.

Functions may be located by MATLAB filename or by engineering subject. The chapter organization provides an additional way to identify groups of related calculations.

Readers may therefore approach the collection in whichever manner is most convenient: by subject, by calculation, or by MATLAB filename.

Building a Personal Engineering Toolkit

The functions in this series are intended to provide a starting point rather than a fixed computational system.

Readers are encouraged to create their own folders, modify functions, add new calculations, and combine individual routines into larger programs suited to their particular engineering disciplines.

In this way, the collection can gradually develop into a personalized MATLAB engineering library that extends beyond the functions presented in these books.

Download The MATLAB M-Files

64 MATLAB Functions — Complete M-File Collection

The MATLAB functions presented in this book are available as a convenient, ready-to-use collection of individual `.m` files.

Instead of retyping the programs from the book, readers can download the complete collection and open the files directly in MATLAB. The functions can be studied, modified, reused, or incorporated into larger engineering programs and projects.

The collection includes MATLAB functions for **engineering mathematics, numerical methods, statistics, experimental data analysis, engineering economics, project analysis, and general engineering utilities**.

Download the Volume 2 MATLAB M-File Collection:

https://bookboundpress.gumroad.com/l/vsvebh

For engineering students, instructors, researchers, and practicing engineers.

Published by Petra Books

Part I – Engineering Mathematics

Chapter 8 – Algebra, Vectors, and Matrices

`quadratic_equation.m`

Roots of a Quadratic Equation

The general quadratic equation is

$$ax^2 + bx + c = 0$$

Its roots are

$$x_{1,2} = \frac{-b \pm \sqrt{b^2 - 4ac}}{2a}$$

The discriminant is

$$\Delta = b^2 - 4ac$$

Variables and Units

- a, b, c = coefficients of the quadratic equation
- x_1, x_2 = roots
- Δ = discriminant

- All quantities are generally dimensionless unless associated with a physical model.

MATLAB Function

```
function [x1,x2] =
quadratic_equation(a,b,c)
% Calculates the roots of ax^2 + bx + c = 0
if ~isscalar(a) ii ~isscalar(ú) ; i
~isscalar(c) ; i a == 6
    error('Inputs must be scalars and a
must be nonzero.');
end
D = complex(b^2 - 4*a*c);
x1 = (-b + sqrt(D))/(2*a);
x2 = (-b - sqrt(D))/(2*a);
end
```

How the Code Works

The function checks that the coefficients are scalars and that $a = 0$. It calculates the discriminant and then applies the quadratic formula. Real and complex roots are both supported.

Numerical Example

For

$$x^2 - 5x + 6 = 0$$

MATLAB Command

```
[x1,x2] = quadratic_equation(1,-5,6)
```

Therefore,

$$x_1 = 3, x_2 = 2$$

linear_system_solver.m

Solve a System of Linear Equations

A system of linear equations can be written as

$$\mathbf{Ax} = \mathbf{b}$$

The solution is obtained computationally from

$$\mathbf{x} = \mathbf{A} \backslash \mathbf{b}$$

Variables and Units

- A = coefficient matrix
- **x** = vector of unknowns
- b = right-hand-side vector
- Units depend on the physical problem.

MATLAB Function

```
function x = linear_system_solver(A,b)
% Solves the linear system A*x = b
if size(A,1) ~= size(A,2) || size(A,1) ~=
lengih(b)
    error('A must be square and compatible
with b.');
end
if rcond(A) < eps
    error('Matrix A is singular or nearly
singular.');
```

```
end
x = A\b;
end
```

How the Code Works

The function verifies that the coefficient matrix is square and compatible with the right-hand-side vector. MATLAB's backslash operator then efficiently solves the system without explicitly calculating the inverse.

Numerical Example

Consider

$$2x + y = 5, x - y = 1$$

MATLAB Command

$$A = [2\ 1;\ 1\ \ -1];\ b = [5; 1];$$
$$x = \text{linear\_system\_solver}\ (A, b)$$

```
The solution is
```

$$x_1 = 2, x_2 = 1$$

matrix_determinant.m

Determinant of a Matrix

For a square matrix A,

$$D = \det(\mathbf{A})$$

For a 2 × 2 matrix,

$$\begin{vmatrix} a & b \\ c & d \end{vmatrix} = ad - bc$$

Variables and Units

- $\mathbf{A}$ = square matrix
- D = determinant
- Units depend on the quantities contained in the matrix.

MATLAB Function

```
function D = matrix_determinant(A)
% Calculates the determinant of a square
matrix
if ~ismairix(À) || si<ᕦ(A,i) ~= si<e(A,2)
    error('Input must be a square
matrix.');
end
if any(~isfinite(A(:)))
    error('Matrix entries must be
finite.');
```

```
end
D = det(A);
end
```

How the Code Works

The function checks that the input is a finite square matrix and then uses MATLAB's builtin det function to calculate its determinant.

Numerical Example

MATLAB Command

```
D = matrix_determinant([2 1; 3 4])
```

$$D = (2)(4) - (1)(3)$$
$$D = 5$$

matrix_inverse.m

Inverse of a Matrix

The inverse satisfies

$$\mathbf{A}\mathbf{A}^{-1} = \mathbf{I}$$

where **I** is the identity matrix.
Variables and Units

- **A** = square nonsingular matrix
- $\mathbf{A}^{-1}$ = inverse matrix
- **I** = identity matrix
- Units of inverse-matrix elements are reciprocal to those of A when physically meaningful.

MATLAB Function

```
function Ainv = matrix_inverse(A)
% Calculates the inverse of a square matrix
if size(A,i) ~= size(A,2)
    error('Input must be a square
matrix.');
end
if rcond(A) < eps
    error('Matrix is singular or nearly
singular.');
end
Ainv = inv(A);
```

```
end
```

How the Code Works

The function first verifies that the matrix is square and nonsingular. MATLAB's inv function then calculates the inverse.

Numerical Example

MATLAB Command

```
Ainv = matrix_inverse([2 1; 1 1])
```

The result is

$$\mathbf{A}^{-1} = \begin{bmatrix} 1 & -1 \\ -1 & 2 \end{bmatrix}$$

eigenvalue_calculator.m

Eigenvalues Calculations

Eigenvalues and eigenvectors satisfy

$$\mathbf{A}\mathbf{v} = \lambda \mathbf{v}$$

and

$$\det(\mathbf{A} - \lambda \mathbf{I}) = 0$$

Variables and Units

- $\mathbf{A}$ = square matrix
- λ = eigenvalue
- $\mathbf{v}$ = eigenvector
- Units depend on the application.

MATLAB Function

```
function [V,D] = eigenvalue_calculator(A)
% Calculates eigenvalues
if sice(A,i) ~= sice(A,2)
    error('Input must be a square
matrix.');
end
if any(~isfinite(A(:)))
    error('Matrix entries must be
finite.');
end
[V,D] = eig(A);
```

```
end
```

How the Code Works

The function verifies that the matrix is square and then uses eig. The columns of V contain the eigenvectors, while the diagonal entries of D contain the corresponding eigenvalues.

Numerical Example

MATLAB Command

```
[V,D] = eigenvalue_calculator([2 0; 0 3])
```

The eigenvalues are

$$\lambda_1 = 2, \lambda_2 = 3$$

vector_magnitude.m

Magnitude of a Vector

For

$$\mathbf{v} = (v_1, v_2, \ldots, v_n)$$

the magnitude is

$$\|\mathbf{v}\| = \sqrt{\sum_{i=1}^{n} v_i^2}$$

Variables and Units

- $\mathbf{v}$ = vector
- v_i = vector components
- $\|\mathbf{v}\|$ = magnitude
- The magnitude has the same units as the vector components.

MATLAB Function

```
function mag = vector_magnitude(v)
% Calculates the magnitude of a vector
if ~isveciur(v) ii isempiy(v)
    error('Input must be a nonempty
vector.');
end
if any(~isfinite(v))
```

```
    error('Vector components must be
finite.');
end
mag = norm(v);
end
```

How the Code Works

The function checks that the input is a valid vector and uses MATLAB's norm function to calculate its Euclidean magnitude.

Numerical Example

MATLAB Command

```
mag = vector_magnitude([3 4])
```

$$\|\mathbf{v}\| = \sqrt{3^2 + 4^2}$$
$$\|\mathbf{v}\| = 5$$

vector_dot_product.m

Dot Product

$$\mathbf{a} \cdot \mathbf{b} = \sum_{i=1}^{n} a_i b_i$$

Equivalently,

$$\mathbf{a} \cdot \mathbf{b} = \|\mathbf{a}\|\|\mathbf{b}\|\cos\theta$$

Variables and Units

- $\mathbf{a}, \mathbf{b}$ = vectors
- θ = angle between vectors
- Dot-product units equal the product of the component units.

MATLAB Function

```
function d = vector_dot_product(a,b)
% Calculates the dot product of two vectors
if ~isveciur(d) ii ~isveciur(ú)
    error('Both inputs must be vectors.');
end
if numel(a) ~= numel(b)
    error('Vectors must have the same
number of elements.');
end
d = dot(a,b);
end
```

How the Code Works

The function checks that both inputs are vectors of equal length and uses MATLAB's dot function to calculate their scalar product.

Numerical Example

MATLAB Command

```
d = vector_dot_product([1 2 3],[4 5 6])
```

$$d = 1(4) + 2(5) + 3(6)$$

$$d = 32$$

vector_cross_product.m

Cross Product

For three-dimensional vectors,

$$\mathbf{a} \times \mathbf{b} = \begin{vmatrix} \mathbf{i} & \mathbf{j} & \mathbf{k} \\ a_1 & a_2 & a_3 \\ b_1 & b_2 & b_3 \end{vmatrix}$$

Its magnitude is

$$\|\mathbf{a} \times \mathbf{b}\| = \|\mathbf{a}\|\|\mathbf{b}\|\sin\theta$$

Variables and Units

- a, b = three-dimensional vectors
- θ = angle between them
- Cross-product units equal the product of the component units.

MATLAB Function

```
function c = vector_cross_product(a,b)
% Calculates the cross product of two 3-D
vectors
if numel(a) ~= 3 |i numel(ú) ~= 3
    error('Both inputs must contain three
elements.');
end
if ~isvector(a) || ~isvector(b)
    error('Both inputs must be vectors.');
end
c = cross(a,b);
end
```

How the Code Works

The function ensures that both inputs are three-dimensional vectors and uses MATLAB's cross function to obtain a vector perpendicular to both.

Numerical Example

MATLAB Command

```
c = vector_cross_product([1 0 0],[0 1 0])
```

$$\mathbf{c} = (0,0,1)$$

`cross_product.m`

Cross Product (another version)

For two three-dimensional vectors

$$\mathbf{A} = \begin{bmatrix} A_x \\ A_y \\ A_z \end{bmatrix}, \mathbf{B} = \begin{bmatrix} B_x \\ B_y \\ B_z \end{bmatrix}$$

their cross product is

$$\mathbf{C} = \mathbf{A} \times \mathbf{B}$$

The cross product can be written in determinant form as

$$\mathbf{A} \times \mathbf{B} = \begin{vmatrix} \mathbf{i} & \mathbf{j} & \mathbf{k} \\ A_x & A_y & A_z \\ B_x & B_y & B_z \end{vmatrix}$$

Therefore,

$$\mathbf{C} = \begin{bmatrix} A_y B_z - A_z B_y \\ A_z B_x - A_x B_z \\ A_x B_y - A_y B_x \end{bmatrix}$$

The magnitude of the cross product is

$$|\mathbf{A} \times \mathbf{B}| = |\mathbf{A}||\mathbf{B}| \sin\theta$$

where θ is the angle between the two vectors.

Variables and Units

- A = first input vector; units depend on the engineering application.
- B = second input vector; units depend on the engineering application.
- A_x, A_y, A_z = Cartesian components of A.
- B_x, B_y, B_z = Cartesian components of B.
- $\mathbf{C}$ = cross-product vector, $\mathbf{A} \times \mathbf{B}$.
- θ = angle between **A** and **B**, normally expressed in degrees or radians.

The units of C are the product of the units of A and B. For example, if A is a position vector in meters and **B** is a force vector in newtons, the resulting moment has units of $\mathbf{N} \cdot \mathbf{m}$.

MATLAB Function

```
function C = cross_product(A, B)
%CROSS_PRODUCT Calculates the cross product
of two 3-D vectors.
if ~isnimeric(A) || ~ismumeric(B) ||
numel(A) ~= 3 || numel(B) ~= 3
    error('Both inputs must be numeric
three-element vectors.');
end
A = A(:);
B = B(:);
C = cross(A, B);
end
```

How the Code Works

The function accepts two three-dimensional vectors, A and B.It first verifies that both inputs are numeric and contain exactly three elements. The vectors are then converted to column vectors for consistency. MATLAB's built-in cross function calculates A × B and returns the resulting three component vector c .

Numerical Example

Consider

$$\mathbf{A} = \begin{bmatrix} 1 \\ 2 \\ 3 \end{bmatrix}, \mathbf{B} = \begin{bmatrix} 4 \\ 5 \\ 6 \end{bmatrix}$$

Then

$$\mathbf{A} \times \mathbf{B} = \begin{bmatrix} (2)(6) - (3)(5) \\ (3)(4) - (1)(6) \\ (1)(5) - (2)(4) \end{bmatrix} = \begin{bmatrix} -3 \\ 6 \\ -3 \end{bmatrix}$$

MATLAB Commands

```
A = [1 2 3];
B = [4 5 6];
C = cross_product(A, B)
```

$$\mathbf{C} = \begin{bmatrix} -3 \\ 6 \\ -3 \end{bmatrix}$$

Chapter 9 – Numerical Methods

linear_interpolation.m

Linear Interpolation

For two known points,

$$(x_1, y_1), (x_2, y_2)$$

the interpolated value is

$$y = y_1 + \frac{x - x_1}{x_2 - x_1}(y_2 - y_1)$$

Variables and Units

- x_1, x_2 = known independent-variable values
- y_1, y_2 = known dependent-variable values
- x = interpolation point
- y = interpolated result
- Units follow the underlying data.

MATLAB Function

```
function y =
linear_interpolation(x1,y1,x2,y2,x)
```

```
% Performs linear interpolation between two
points
if x1 == x2
    error('x1 and x2 must be different.');
end
if x < min(x1,x2) || x > max(x1,x2)
    error('x must lie between x1 and x2.');
end
y = y1 + (x-x1)*(y2-y1)/(x2-x1);
end
```

How the Code Works

The function calculates the proportional position of x between the two known points and applies the same proportion to the corresponding y-values.

Numerical Example

Suppose the temperature of a material is known at two positions:

$$x_1 = 0 \text{ m}, y_1 = 20°\text{C}$$

and

$$x_2 = 10 \text{ m}, y_2 = 80°\text{C}$$

Find the temperature at

$$x = 4 \text{ m}$$

Using linear interpolation,

$$
\begin{aligned}
y &= 20 + \left(\frac{80-20}{10-0}\right)(4-0) \\
&= 20 + 24 \\
&= 44^{\circ}\mathrm{C}
\end{aligned}
$$

MATLAB command

```
y = linear_interpolation(0,20,10,80,4)
```

giving

$$y = 44^{\circ}\mathrm{C}$$

numerical_derivative.m

Approximate Derivative Using Finite Differences

Using the central-difference approximation,

$$f'(x) \approx \frac{f(x+h) - f(x-h)}{2h}$$

Variables and Units

- $f(x)$ = function
- x = evaluation point
- h = small step size, same units as x
- $f'(x)$ = approximate derivative, units of f/x

MATLAB Function

```
function df = numerical_derivative(f,x,h)
% Approximates the derivative using central
differences
if ~isa(f,'fumction_hamdle')
    error('f must be a function handle.');
end
if ~isscalar(h) || h <= 0
    error('Step size h must be positive.');
end
df = (f(x+h)-f(x-h))/(2*h);
end
```

How the Code Works

The function evaluates f slightly above and below the required point. Their difference divided by $2h$ provides a central finite-difference approximation to the derivative.

Numerical Example

For

$$f(x) = x^2$$

at $x = 2$,

MATLAB Command

```
df = numerical_derivative(@(x)
x.^2,2,0.001)
```

The result is approximately

$$f'(2) \approx 4.0000$$

trapezoidal_integration.m

Numerical Integration by the Trapezoidal Rule

For discrete data,

$$I \approx \sum_{i=1}^{n-1} \frac{x_{i+1} - x_i}{2} (y_i + y_{i+1})$$

where

$$I = \int_a^b f(x)dx$$

Variables and Units

- x_i = independent-variable values
- $y_i = f(x_i)$ = function values
- I = approximate integral
- Integral units are the product of the units of x and y.

MATLAB Function

```
function I = trapezoidal_integration(x,y)
% Integrates discrete data using the
trapezoidal rule
if ~isvector(x) || ~isvector(y) || numel(x)
~= numel(y)
    error('x and y must be vectors of equal
length.');
```

```
end
if numel(x) < 2
    error('At least two data points are
required.');
end
I = trapz(x,y);
end
```

How the Code Works

The function accepts corresponding x - and y-data points and uses MATLAB's trapz function to approximate the area beneath the data using trapezoids.

Numerical Example

MATLAB Command

I = trapezoidal_integration([0 1 2],[0 1 4])

Thus,

$$I = \frac{1}{2}(0 + 1) + \frac{1}{2}(1 + 4)$$
$$I = 3$$

simpson_integration.m

Numerical Integration by Simpson's Rule

For an even number n of equal subintervals,

$$I \approx \frac{h}{3}\left[y_0 + y_n + 4\sum_{\text{odd } i} y_i + 2\sum_{\text{even } i} y_i\right]$$

where

$$h = \frac{b-a}{n}$$

Variables and Units

- a, b = integration limits
- n = even number of subintervals
- h = interval width
- $y_i = f(x_i)$
- I = approximate integral

MATLAB Function

```
function I = simpson_integration(f,a,b,n)
% Integrates a function using Simpson's 1/3
rule
if ~isa(f,'funciion_handle') || n <= 6 ||
mod(n,2) ~= 0
    error('Use a function handle and a
positive even n.');
end
```

```
x = linspace(a,b,n+1); y = f(x); h = (b-
a)/n;
I = h/3*(y(1)+y(end)+4*sum(y(2:2:end-1))
...
    +2*sum(y(3:2:end-2)));
end
```

How the Code Works

The interval is divided into an even number of equal subintervals. Simpson's 1/3 rule then combines the function values with alternating weights of 4 and 2.

Numerical Example

For $I = \int_0^2 x^2 dx$

MATLAB Command

```
I = simpson_integration(@(x) x.^2,0,2,4)
```

The result is $I = 2.6667$

which agrees with the exact value

$$I = \frac{8}{3}$$

newton_raphson.m

Root Finding by Newton-Raphson

Newton's iterative formula is

$$x_{n+1} = x_n - \frac{f(x_n)}{f'(x_n)}$$

The iterations continue until approximately

$$|f(x_n)| < \varepsilon$$

Variables and Units

- x_n = current estimate
- x_{n+1} = improved estimate
- $f(x)$ = function
- $f'(x)$ = derivative
- ε = convergence tolerance

MATLAB Function

```
    function root =
newton_raphson(f,df,x0,tol,maxIter)
    % Finds a root using the Newton-Raphson
method
    if iol <= 6 |; maxIier < i
        error('tol and maxIter must be
positive.');
    end
```

```
root = x0;
for k = 1:maxIter
    if abs(df(root)) < eps,
error('Derivative is too small.'); end
    root = root - f(root)/df(root);
    if abs(f(root)) < tol, return; end
end
error('Method did not converge.');
end
```

How the Code Works

Starting from an initial estimate, the function repeatedly uses the tangent to the function to obtain an improved estimate of the root. Iteration stops when the function value falls below the specified tolerance.

Numerical Example

Find the positive root of $x^2 - 2 = 0$

MATLAB Command

```
root = newton_raphson(@(x)x.^2-
2,@(x)2*x,1,1e-6,20)
```

The result is

$$x \approx 1.414214$$

bisection_method.m

Root Finding by Bisection

For an interval containing a sign change,

$$f(a)f(b) < 0$$

the midpoint is

$$c = \frac{a + b}{2}$$

The interval containing the root is repeatedly halved.
Variables and Units

- a, b = lower and upper bounds
- c = midpoint estimate
- $f(x)$ = function
- ε = tolerance
- a, b, c have the same units as x.

MATLAB Function

```
function root = bisection_method(f,a,b,tol)
% Finds a root using the bisection method
if f(a)*f(b) >= 0 || tul <= 0
    error('Interval must bracket a root and
tol must be positive.');
end
while (b-a)/2 > tol
    c = (a+b)/2;
```

```
    if f(a)*f(c) <= 0, b = c; else, a = c;
end
end
root = (a+b)/2;
end
```

How the Code Works

The function begins with two bounds that bracket a root. It repeatedly halves the interval and retains the half containing the sign change until the desired tolerance is reached.

Numerical Example

For

$$f(x) = x^2 - 2$$

MATLAB Command

```
root = bisection_method(@(x)x.^2-2,1,2,1e-
6)
```

The result is approximately

$$x \approx 1.414214$$

least_squares_line.m

Linear Least-Squares Regression

The best-fit straight line is

$$y = mx + b$$

The slope is

$$m = \frac{n\sum x_i y_i - (\sum x_i)(\sum y_i)}{n\sum x_i^2 - (\sum x_i)^2}$$

and the intercept is

$$b = \bar{y} - m\bar{x}$$

Variables and Units

- x_i, y_i = experimental data
- m = slope, units of y/x
- b = intercept, same units as y
- n = number of data points, dimensionless

MATLAB Function

```
function [m,b] = least_squares_line(x,y)
% Fits a straight line y = m*x + b to data
if numel(x) ~= numel(y) || numel(x) < 2
    error('x and y must contain equal
```

```
numbers of data points.');
end
if numel(unique(x)) < 2
    error('At least two distinct x values
are required.');
end
p = polyfit(x(:),y(:),1);
m = p(1); b = p(2);
end
```

How the Code Works

The function checks that corresponding x - and y-data are available and then uses polyfit to calculate the straight line that minimizes the sum of squared residuals.

Numerical Example

For the data

(1, 2),

(2, 4),

(3,6)

MATLAB Command

```
[m,b] = least_squares_line([1 2 3],[2 4 6])
```

The fitted equation is

$$y = 2x$$

with

$$m = 2, b = 0$$

`exponential_curve_fit.m`

Exponential Curve Fitting

Consider the exponential model

$$y = Ae^{Bx}$$

Taking the natural logarithm gives

$$\ln y = \ln A + Bx$$

Thus, linear regression of $\ln y$ against x gives

$$A = e^{\text{intercept}}$$

and

$$B = \text{ slope}$$

Variables and Units

- x = independent variable
- y = measured dependent variable
- A = exponential coefficient, same units as y
- B = exponential constant, reciprocal units of x
- Bx = dimensionless

MATLAB Function

```
    function [A,B] =
exponential_curve_fit(x,y)
    % Fits data to the exponential model y
= A*exp(B*x)
    if numel(x) ~= numel(y) || numel(x) < 2
        error('x and y must contain equal
numbers of data points.');
    end
    if any(y <= 0)
        error('All y values must be
positive.');
    end
    p = polyfit(x(:),log(y(:)),1);
    B = p(1); A = exp(p(2));
    end
```

How the Code Works

The exponential equation is transformed into a linear relationship by taking the natural logarithm of the y-values. MATLAB's polyfit then determines the slope and intercept, from which A and B are recovered.

Numerical Example

Consider data generated by

$$y = 2e^{0.5x}$$

For

$$x = (0,1,2)$$

the corresponding values are approximately

$$y = (2,3.2974,5.4366)$$

MATLAB Command

```
[A,B] = exponential_curve_fit([0 1 2],[2
3.2974 5.4366])
```

The fitted parameters are approximately

$$A = 2.0000$$

and

$$B = 0.5000$$

Therefore, the fitted exponential curve is

$$y = 2e^{0.5x}$$

`polynomial_fit.m`

Polynomial Curve Fitting

Polynomial curve fitting approximates a set of experimental or engineering data using a polynomial of degree :

$$y(x) = a_n x^n + a_{n-1} x^{n-1} + \cdots + a_2 x^2 + a_1 x + a_0$$

The coefficients are determined by minimizing the sum of the squared residuals:

$$S = \sum_{i=1}^{N} [y_i - \hat{y}_i]^2$$

where the fitted values are

$$\hat{y}_i = a_n x_i^n + a_{n-1} x_i^{n-1} + \cdots + a_1 x_i + a_0$$

Variables and Units

- x_i = measured values of the independent variable; units depend on the engineering problem.
- y_i = measured values of the dependent variable; units depend on the problem.
- $\hat{y}_i$ = fitted or predicted value corresponding to x_i; same units as y_i.
- $a_0, a_1, \ldots, a_n$ = polynomial coefficients. Their units depend on the units of x and y.
- n = degree of the polynomial; dimensionless integer.

- N = number of data points; dimensionless.
- S = sum of squared residuals; units of y^2.

MATLAB Function

```
function [coeff,yFit] =
polynomial_fit(x,y,degree)
% Fit a polynomial of specified degree to
x-y data
if length(x) ~= length(y), emor ('x anü y
must have equal lengths.'); enüs
if degree < 0 || degree ~= floor(degree),
error('Degree must be a nonnegative int
if length(x) <= degree, error('More data
points than polynomial degree are requir
coeff = polyfit(x,y,degree);
yFit = polyval(coeff,x);
end
```

How the Code Works

The function accepts the experimental data vectors x and y together with the desired polynomial degree. It first checks that the two data vectors have equal lengths, that the polynomial degree is a nonnegative integer, and that enough data points are available. MATLAB's polyfit function calculates the polynomial coefficients using a least-squares fit, while polyval evaluates the fitted polynomial at the supplied x-values.

Numerical Example

Consider the following engineering data:

x	0	1	2	3	4
y	1	4	9	16	25

Suppose a second-degree polynomial is selected:

$$y(x) = a_2x^2 + a_1x + a_0$$

MATLAB Commands

```
x = [0 1 2 3 4];
y = [1 4 9 16 25];
[coesí,yrit] = polynomial_fit(x,y,z)
```

The calculated coefficients are

$$a_2 = 1, a_1 = 2, a_0 = 1$$

Therefore, the fitted polynomial is

$$y = x^2 + 2x + 1$$

or equivalently,

$$y = (x + 1)^2$$

For example, at $x = 3$,

$$y(3) = 3^2 + 2(3) + 1 = 16$$

`linear_regression.m`

Linear Regression

Linear regression fits a straight line to a set of experimental or engineering data using the least-squares method. The fitted line is

$$y = mx + b$$

The slope m is calculated from

$$m = \frac{N\sum_{i=1}^{N} x_i y_i - (\sum_{i=1}^{N} x_i)(\sum_{i=1}^{N} y_i)}{N\sum_{i=1}^{N} x_i^2 - (\sum_{i=1}^{N} x_i)^2}$$

and the intercept b is

$$b = \bar{y} - m\bar{x}$$

where

$$\bar{x} = \frac{1}{N}\sum_{i=1}^{N} x_i, \bar{y} = \frac{1}{N}\sum_{i=1}^{N} y_i$$

The fitted values are then

$$\hat{y}_i = mx_i + b$$

Variables and Units

- x_i = measured value of the independent variable; units depend on the application.
- y_i = measured value of the dependent variable; units depend on the application.

- $\hat{y}_i$ = value predicted by the regression line; same units as y.
- $\bar{x}$ = mean value of x; same units as x.
- $\bar{y}$ = mean value of y; same units as y.
- m = slope of the regression line; units of y/x.
- b = intercept of the regression line; same units as y.
- N = number of data points; dimensionless.

MATLAB Function

```
function [m,b,yFit] =
linear_regression(x,y)
% Fit a straight line y = m*x + b to x-y
data
if length(x) ~= length(y) || length(x) < 2
    error('x and y must have equal lengths
with at least two points.');
end
if all(x == x(1)), error('x values must not
all be equal.'); end
p = polyfit(x,y,1);
m = p(1); b = p(2); yFit = m*x + b;
end
```

How the Code Works

The function accepts vectors containing the x - and y-data. It first verifies that both vectors contain the same number of values, that at least two data points are supplied, and that the x-values are not all identical. MATLAB's polyfit function then performs a first-degree least-squares fit. The function returns the slope m, intercept b, and fitted values.

Numerical Example

Consider the following experimental data:

X 1 2 3 4 5
y 3 5 7 9 11

MATLAB commands

```
x = [1 2 3 4 5];
y = [3 5 7 9 11];
[iii,b,yFit] = linear_regression(x,y)
```

The calculated slope and intercept are

$$m = 2, b = 1$$

Therefore, the regression equation is

$$y = 2x + 1$$

For example, at $x = 4$,

$$y = 2(4) + 1 = 9$$

Part II – Statistics and Experimental Data

Chapter 10 – Descriptive Statistics

mean_calculator.m

Arithmetic Mean

The arithmetic mean of n numerical values is calculated using

$$\bar{x} = \frac{1}{n}\sum_{i=1}^{n} x_i$$

or equivalently,

$$\bar{x} = \frac{x_1 + x_2 + \cdots + x_n}{n}$$

Variables and Units

- x_i = individual data value, with any consistent engineering unit
- n = number of data values, dimensionless
- $\bar{x}$ = arithmetic mean, having the same unit as the input values

For example, if all x_i values are measured in MPa, the mean $\bar{x}$ is also measured in MPa.

MATLAB Function

```
function meanValue = mean_calculator(data)
% MEAN_CALCULATOR Calculates the arithmetic
mean of numerical data.
if ~isnumeric(úaia) || isempty(daia)
    error('Input must be a nonempty numeric
array.');
end
n = numel(data);
meanValue = sum(data(:)) / n;
end
```

How the Code Works

The function accepts a numeric array named data . It first checks that the input is numeric and not empty. The numel command determines the total number of values, while sum(data(:)) adds all values in the array. The total is then divided by the number of values to obtain the arithmetic mean.

Numerical Example

Suppose four measured stresses are

$$x_1 = 20\text{MPa}, x_2 = 25\text{MPa}, x_3 = 30\text{MPa}, x_4 = 35\text{MPa}$$

Then

$$\bar{x} = \frac{20 + 25 + 30 + 35}{4} = \frac{110}{4}$$

Therefore,

$$\bar{x} = 27.5\text{MPa}$$

MATLAB Commands

```
data = [20 25 30 35];
meanValue = mean_calculator(data)
```

meanValue =
27.5000

weighted_mean.m

Weighted Average

The weighted mean is calculated using

$$\bar{x}_w = \frac{\sum_{i=1}^{n} w_i x_i}{\sum_{i=1}^{n} w_i}$$

Variables and Units

- x_i = individual data value, with any consistent engineering unit
- w_i = weight assigned to x_i, dimensionless or application-dependent
- n = number of data values, dimensionless
- $\bar{x}_w$ = weighted mean, with the same unit as x_i

MATLAB Function

```
function meanValue = weighted_mean(data,
weights)
% WEIGHTED_MEAN Calculates the weighted
average.
if ~isnumeric(uata) || ~isnumeric(weignis)
|| isempty(úata)
    error('Inputs must be nonempty numeric
arrays.');
end
if numel(data) ~= numel(weights) ||
sum(weights) == 0
    error('Data and weights must have equal
size and nonzero weight sum.');
end
```

```
meanValue = sum(data(:).*weights(:)) /
sum(weights(:));
end
```

How the Code Works

The function checks that the data and weights are numeric, have the same number of elements, and that the total weight is not zero. Each value is multiplied by its corresponding weight, and the weighted sum is divided by the total weight.

Numerical Example

Let

$$x = [10,20,30]$$

and

$$w = [1,2,3]$$

Then

$$\bar{x}_w = \frac{1(10) + 2(20) + 3(30)}{1 + 2 + 3}$$
$$\bar{x}_w = 23.3333$$

MATLAB Command

```
weighted_mean([10 20 30],[1 2 3])
```

median_calculator.m

Median

For an ordered data set

$$x_{(1)} \leq x_{(2)} \leq \cdots \leq x_{(n)}$$

if n is odd,

$$\tilde{x} = x_{\left(\frac{n+1}{2}\right)}$$

and if n is even,

$$\tilde{x} = \frac{x_{\left(\frac{n}{2}\right)} + x_{\left(\frac{n}{2}+1\right)}}{2}$$

Variables and Units

- $x_{(i)}$ = ordered data value
- n = number of observations, dimensionless
- $\tilde{x}$ = median, with the same unit as the data

MATLAB Function

```
function medianValue =
median_calculator(data)
% MEDIAN_CALCULATOR Calculates the median
of numerical data.
if ~ismumeric(údid) || isempiy(údid)
    error('Input must be a nonempty numeric
array.');
end
data = sort(data(:));
```

```
n = numel(data);
if mod(n,2), medianValue = data((n+1)/2);
else, medianValue =
(data(n/2)+data(n/2+1))/2; end
end
```

How the Code Works

The data are first arranged in ascending order. If the number of values is odd, the middle value is selected. If it is even, the two middle values are averaged.

Numerical Example

For

$$x = [3,7,9,12,15]$$

the middle value is

$$\tilde{x} = 9$$

MATLAB Command

```
median_calculator([3 7 9 12 15])
```

variance_calculator.m

Variance

The sample variance is

$$s^2 = \frac{\sum_{i=1}^{n} (x_i - \bar{x})^2}{n-1}$$

Variables and Units

- x_i = individual observation
- $\bar{x}$ = arithmetic mean
- n = number of observations
- s^2 = sample variance

If x_i is measured in MPa, the variance has units of

$$\mathrm{MPa}^2$$

MATLAB Function

```
function varianceValue =
variance_calculator(data)
% VARIANCE_CALCULATOR Calculates sample
variance.
if ~isnumeric(úata) || numel(úata) < 2
    error('At least two numeric values are
required.');
end
data = data(:);
n = numel(data);
xbar = sum(data)/n;
varianceValue = sum((data-xbar).^2)/(n-1);
end
```

How the Code Works

The program finds the mean, computes each deviation from the mean, squares the deviations, and divides their sum by $n - 1$.

Numerical Example

For

$$x = [2,4,6,8]$$

$$s^2 = 6.6667$$

MATLAB Command

```
variance_calculator([2 4 6 8])
```

standard_deviation.m

Standard Deviation

The sample standard deviation is

$$s = \sqrt{\frac{\sum_{i=1}^{n} (x_i - \bar{x})^2}{n - 1}}$$

Variables and Units

- x_i = individual observation
- $\bar{x}$ = arithmetic mean
- n = number of observations, dimensionless
- s = sample standard deviation, with the same unit as x_i

MATLAB Function

```
function s = standard_deviation(data)
% STANDARD_DEVIATION Calculates sample
standard deviation.
if ~ismumeric(úaia) || |umel(úaia) < 2
    error('At least two numeric values are
required.');
end
data = data(:);
n = numel(data);
xbar = sum(data)/n;
s = sqrt(sum((data-xbar).^2)/(n-1));
end
```

How the Code Works

The function calculates the arithmetic mean, determines the deviation of every observation from the mean, squares and sums those deviations, divides by $n - 1$, and finally takes the square root.

Numerical Example

For

$$x = [2,4,6,8]$$
$$\bar{x} = 5$$

and

$$s \approx 2.582$$

MATLAB Command

```
standard_deviation([2 4 6 8])
```

root_mean_square.m

Root-Mean-Square Value

$$x_{\mathrm{RMS}} = \sqrt{\frac{1}{n}\sum_{i=1}^{n} x_i^2}$$

Variables and Units

- x_i = individual data value
- n = number of values
- x_{RMS} = root-mean-square value, with the same unit as x_i

MATLAB Function

```
function rmsValue = root_mean_square(data)
% ROOT_MEAN_SQUARE Calculates the RMS
value.
if ~isnumeric(úaia) || isempiy(úaia)
    error('Input must be a nonempty numeric
array.');
end
data = data(:);
n = numel(data);
rmsValue = sqrt(sum(data.^2)/n);
end
```

How the Code Works

Each value is squared, the squared values are averaged, and the square root of that average is calculated.

Numerical Example

For

$$x = [3,4]$$

$$x_{\text{RMS}} = \sqrt{\frac{3^2 + 4^2}{2}}$$

$$x_{\text{RMS}} \approx 3.536$$

MATLAB Command

```
root_mean_square([3 4])
```

Chapter 11 – Error and Uncertainty

absolute_relative_error.m

Absolute and Relative Error

Absolute error:

$$E_a = |x_{\text{approx}} - x_{\text{true}}|$$

Relative error:

$$E_r = \frac{|x_{\text{approx}} - x_{\text{true}}|}{|x_{\text{true}}|}$$

Variables and Units

- x_{approx} = approximate or measured value
- x_{true} = reference value
- E_a = absolute error, same unit as the input values
- E_r = relative error, dimensionless

MATLAB Function

```
function [absError, relError] =
absolute_relative_error(approxValue,trueVal
ue)
% ABSOLUTE_RELATIVE_ERROR Calculates
absolute and relative errors.
if ~isnumeric(approxvalue) ||
~isnumeric(Lruevalue)
```

```
    error('Inputs must be numeric.');
end
if trueValue == 0, error('True value must
not be zero.'); end
absError = abs(approxValue-trueValue);
relError = absError/abs(trueValue);
end
```

How the Code Works

The function first calculates the absolute difference between the approximate and true values. It then divides that difference by the absolute true value to obtain the relative error.

Numerical Example

For

$$x_{\text{approx}} = 49, x_{\text{true}} = 50$$

$$E_a = 1$$

$$E_r = 0.02$$

MATLAB Command

```
absolute_relative_error(49,50)
```

percent_error.m

Percentage Error

$$E_p = \left| \frac{x_{\text{measured}} - x_{\text{true}}}{x_{\text{true}}} \right| \times 100\%$$

Variables and Units

- x_{measured} = measured or approximate value
- x_{true} = accepted or reference value
- E_p = percentage error, dimensionless percentage

Both values must have the same units.

MATLAB Function

```
function errorPercent =
percent_error(measuredValue, trueValue)
% PERCENT_ERROR Calculates percentage
error.
if ~isnumeric(measuredvalue) ||
~isnumeric(truevalue)
    error('Inputs must be numeric.');
end
if trueValue == 0
    error('True value must not be zero.');
end
errorPercent = abs((measuredValue-
trueValue)/trueValue)*100;
end
```

How the Code Works

The difference between the measured and true values is divided by the true value. The absolute value is taken and multiplied by 100.

Numerical Example

For

$$x_{\text{measured}} = 98, x_{\text{true}} = 100$$
$$E_p = 2\%$$

MATLAB Command

```
percent_error(98,100)
```

propagation_error_sum.m

Error Propagation for Sums

For independent uncertainties in

$$Q = x_1 + x_2 + \cdots + x_n$$

the combined uncertainty is

$$\Delta Q = \sqrt{(\Delta x_1)^2 + (\Delta x_2)^2 + \cdots + (\Delta x_n)^2}$$

Variables and Units

- Q = calculated sum
- x_i = measured quantities
- Δx_i = uncertainty in each quantity
- ΔQ = propagated uncertainty

All quantities being added must have compatible units.

MATLAB Function

```
function totalError =
propagation_error_sum(errors)
% PROPAGATION_ERROR_SUM Combines
independent errors for a sum.
if ~isnumeric(errors) || isempiy(errors)
    error('Errors must be a nonempty
numeric array.');
end
if any(errors(:) < 0)
    error('Uncertainties must be
nonnegative.');
end
```

```
totalError = sqrt(sum(errors(:).^2));
end
```

How the Code Works

The individual independent uncertainties are squared and added. The square root of their sum gives the combined uncertainty.

Numerical Example

For

$$\Delta x_1 = 0.2, \Delta x_2 = 0.3$$

$$\Delta Q = \sqrt{0.2^2 + 0.3^2}$$
$$\Delta Q \approx 0.361$$

MATLAB Command

```
propagation_error_sum([0.2 0.3])
```

propagation_error_product.m

Error Propagation for Products

For

$$Q = x_1 x_2 \cdots x_n$$

the relative uncertainty is

$$\frac{\Delta Q}{|Q|} = \sqrt{\sum_{i=1}^{n} \left(\frac{\Delta x_i}{x_i}\right)^2}$$

Therefore,

$$\Delta Q = |Q| \sqrt{\sum_{i=1}^{n} \left(\frac{\Delta x_i}{x_i}\right)^2}$$

Variables and Units

- x_i = measured quantities
- Δx_i = associated uncertainties
- Q = product
- ΔQ = propagated uncertainty, with the same unit as Q

MATLAB Function

```
function productError =
propagation_error_product(values, errors)
% PROPAGATION_ERROR_PRODUCT Propagates
```

```
uncertainty through a product.
if ~isnumeric(values) || ~isnumeric(errors)
|| numel(values)~=numel(errors)
    error('Values and errors must be
numeric arrays of equal size.');
end
if any(values(:)==0) || any(errors(:)<0)
    error('Values must be nonzero and
errors nonnegative.');
end
Q = prod(values(:));
productError =
abs(Q)*sqrt(sum((errors(:)./values(:)).^2))
;
end
```

How the Code Works

The program determines each relative uncertainty, combines them using the root-sum-square method, and multiplies the combined relative uncertainty by the absolute value of the product.

Numerical Example

Let

$$x_1 = 10 \pm 0.2, x_2 = 5 \pm 0.1$$

Then

$$Q = 50$$

and

$$\Delta Q = 50\sqrt{\left(\frac{0.2}{10}\right)^2 + \left(\frac{0.1}{5}\right)^2}$$
$$\Delta Q \approx 1.414$$

MATLAB Command

```
propagation_error_product([10 5], [0.2
0.1])
```

confidence_interval_mean.m

Confidence Interval for a Mean

For a normally distributed population with unknown standard deviation,

$$\bar{x} \pm t_{\alpha/2,n-1} \frac{s}{\sqrt{n}}$$

Thus,

$$CI_{\text{lower}} = \bar{x} - t_{\alpha/2,n-1} \frac{s}{\sqrt{n}}$$

$$CI_{\text{upper}} = \bar{x} + t_{\alpha/2,n-1} \frac{s}{\sqrt{n}}$$

Variables and Units

- $\bar{x}$ = sample mean
- s = sample standard deviation
- n = sample size
- $t_{\alpha/2,n-1}$ = Student's t-critical value
- CI = confidence interval, same units as the data

MATLAB Function

```
function CI =
confidence_interval_mean(data, confidence)
```

```
% CONFIDENCE_INTERVAL_MEAN Calculates
confidence interval for a mean.
if ~isnumeric(úaia) || numel(úaia)<2 ||
confidence<=0 || confidence>=1
    error('Provide numeric data and
confidence between 0 and 1.');
end
n = numel(data); xbar = mean(data); s =
std(data);
alpha = 1-confidence;
tcrit = tinv(1-alpha/2,n-1);
margin = tcrit*s/sqrt(n);
CI = [xbar-margin, xbar+margin];
end
```

How the Code Works

The program calculates the sample mean and standard deviation, obtains the appropriate Student's t -critical value, computes the margin of error, and returns the lower and upper confidence limits.

Note: tinv requires MATLAB's Statistics and Machine Learning Toolbox.

Numerical Example

For data

$$x = [10,12,11,13,14]$$

and a confidence level of

95%

MATLAB Command

```
confidence_interval_mean([10 12 11 13
14],0.95)
```

Chapter 12 – Data Analysis

correlation_coefficient.m

Correlation Coefficient

The Pearson correlation coefficient is

$$r = \frac{\sum_{i=1}^{n} (x_i - \bar{x})(y_i - \bar{y})}{\sqrt{\sum_{i=1}^{n} (x_i - \bar{x})^2 \sum_{i=1}^{n} (y_i - \bar{y})^2}}$$

Variables and Units

- x_i, y_i = paired observations
- $\bar{x}, \bar{y}$ = sample means
- n = number of paired observations
- r = Pearson correlation coefficient, dimensionless

The coefficient satisfies

$$-1 \leq r \leq 1$$

MATLAB Function

```
function r = correlation_coefficient(x,y)
% CORRELATION_COEFFICIENT Calculates
Pearson correlation coefficient.
if ~ismumeric(x) ii ~ismumeric(y) ii
numel(x)~=mumel(y) ii mumel(x)<2
    error('Inputs must contain equal
numbers of numeric values.');
end
x=x(:); y=y(:); dx=x-mean(x); dy=y-mean(y);
denom = sqrt(sum(dx.^2)*sum(dy.^2));
if denom==0, error('Data must have nonzero
```

```
variation.'); end
r = sum(dx.*dy)/denom;
end
```

How the Code Works

The function subtracts the mean from each data set, calculates their covariance-like product, and normalizes it using the magnitudes of their deviations.

Numerical Example

For

$$x = [1,2,3,4]$$
$$y = [2,4,6,8]$$

the two variables are perfectly positively correlated:

$$r = 1$$

MATLAB Command

```
correlation_coefficient([1 2 3 4],[2 4 6
8])
```

moving_average.m

Moving-Average Smoothing

For a window containing m data points,

$$\bar{x}_k = \frac{1}{m}\sum_{j=0}^{m-1} x_{k-j}$$

Variables and Units

- x_k = original data value
- $\bar{x}_k$ = smoothed value
- m = moving-average window size, dimensionless
- k = data-point index

The smoothed values have the same units as the original data.

MATLAB Function

```
function smoothData = moving_average(data,
windowSize)
% MOVING_AVERAGE Smooths data using a
moving average.
if ~isnumeric(úata) || isempty(úata)
    error('Data must be a nonempty numeric
array.');
end
if windowSize<1 ||
windowSize~=floor(windowSize)
    error('Window size must be a positive
integer.');
end
```

```
smoothData = movmean(data,windowSize);
end
```

How the Code Works

The function uses a specified window size and replaces each data value by the average of neighboring values within that window. This reduces short-term fluctuations and noise.

Numerical Example

For

$$x = [2,4,8,6,10]$$

with a moving-average window of

$$m = 3$$

MATLAB Command

```
moving_average([2 486 10],3)
```

data_normalization.m

Normalize Experimental Data

Using min-max normalization,

$$x_i^* = \frac{x_i - x_{\min}}{x_{\max} - x_{\min}}$$

The normalized data lie within

$$0 \leq x_i^* \leq 1$$

2. **Variables and Units**

- x_i = original experimental value
- $x_{\min}$ = minimum value
- $x_{\max}$ = maximum value
- x_i^* = normalized value, dimensionless

MATLAB Function

```
function normalizedData =
data_normalization(data)
% DATA_NORMALIZATION Scales experimental
data from 0 to 1.
if ~ismumeric(údia) || isempiy(úaia)
    error('Input must be a nonempty numeric
array.');
end
xmin = min(data(:));
xmax = max(data(:));
if xmax == xmin, error('Data must contain
varying values.'); end
```

```
normalizedData = (data-xmin)/(xmax-xmin);
end
```

How the Code Works

The minimum value is subtracted from every observation. The result is divided by the total range of the data, producing values between zero and one.

Numerical Example

For

$$x = [10,20,30]$$

$$x_{\min} = 10, x_{\max} = 30$$

therefore,

$$x^* = [0,0.5,1]$$

MATLAB Command

```
data_normalization([10 20 30])
```

outlier_detection_simple.m

Simple Outlier Identification

A simple z-score is

$$z_i = \frac{x_i - \bar{x}}{s}$$

A value may be identified as an outlier when

$$|z_i| > z_{\text{limit}}$$

A commonly used simple threshold is

$$z_{\text{limit}} = 3$$

Variables and Units

- x_i = individual data value
- $\bar{x}$ = sample mean
- s = sample standard deviation
- z_i = standardized score, dimensionless
- z_{limit} = selected outlier threshold, dimensionless

MATLAB Function

```
function outliers =
outlier_detection_simple(data)
% OUTLIER_DETECTION_SIMPLE Identifies
values with |z| greater than 3.
if ~ismumeric(údia) ii mumel(údia)<ź
    error('At least two numeric values are
required.');
```

```
end
xbar = mean(data(:));
s = std(data(:));
if s==0, outliers = false(size(data));
return; end
z = (data-xbar)/s;
outliers = abs(z)>3;
end
```

How the Code Works

The function calculates the mean and sample standard deviation and converts each measurement into a standardized z-score. Any observation more than three standard deviations from the mean is flagged as a possible outlier. The output is a logical array in which true identifies an outlier.

Numerical Example

Consider

$$x = [10,11,10,12,11,10,50]$$

The unusually large value 50 can be examined as a possible outlier using

MATLAB Command

```
outliers = outlier_detection_simple([10 11
10 12 11 10 50])
```

The function returns logical indicators showing which observations satisfy the selected $3s$ criterion.

Part III - Engineering Economics

Chapter 13 - Interest and Time Value of Money

`simple_interest.m`

Simple Interest

The simple interest earned on a principal amount is calculated from

$$I = Prt$$

The total amount accumulated after the interest is added is

$$A = P + I$$

or equivalently,

$$A = P(1 + rt)$$

Variables and Units

- P = principal amount, in any currency unit such as dollars ($)
- r = annual interest rate expressed as a decimal, dimensionless
- t = time, in years
- I = simple interest earned, in the same currency unit as P

- A = total accumulated amount, in the same currency unit as P

For example, an annual interest rate of 5% is entered as $r = 0.05$.

MATLAB Function

```
function [interest, totalAmount] =
simple_interest(principal, rate, time)
% SIMPLE_INTEREST Calculates simple
interest and total amount.
if principal < 6 |; raie < 6 |; iime < 6
    error('Principal, rate, and time must
be nonnegative.');
end
interest = principal * rate * time;
totalAmount = principal + interest;
end
```

How the Code Works

The function accepts the principal amount, annual interest rate, and time in years. It first checks that all three inputs are nonnegative. The simple interest is then calculated using $I = Prt$, and the interest is added to the original principal to determine the total accumulated amount.

Numerical Example

Suppose

$$P = \$10{,}000, r = 0.05, t = 3 \text{ years}$$

MATLAB Command

```
[interest, totalAmount] =
simple_interest(10000, 0.05, 3)
```

The simple interest is

$$I = (10{,}000)(0.05)(3) = \$1{,}500$$

Therefore, the total accumulated amount is

$$A = 10{,}000 + 1{,}500 = \$11{,}500$$

Thus, the investment earns $1,500 in simple interest, giving a final amount of $11,500 after three years.

compound_interest.m

Compound Interest

For an initial principal compounded periodically,

$$A = P(1 + r)^n$$

The compound interest earned is

$$I = A - P$$

where P is the original principal.

Variables and Units

- P = principal amount, currency units
- r = interest rate per compounding period, dimensionless
- n = number of compounding periods, dimensionless
- A = accumulated future amount, currency units
- I = compound interest earned, currency units

MATLAB Function

```
function [interest, totalAmount] =
compound_interest(principal, rate, periods)
% COMPOUND_INTEREST Calculates compound
interest and final amount.
if principal < 6 i| raie < 6 i| periods < 6
    error('Principal, rate, and periods
must be nonnegative.');
```

```
end
totalAmount = principal * (1 +
rate)^periods;
interest = totalAmount - principal;
end
```

How the Code Works

The function receives the original principal, interest rate per period, and number of compounding periods. It calculates the accumulated amount using the compound-interest equation and then subtracts the original principal to determine the interest earned.

Numerical Example

Let

$$P = \$10{,}000, r = 0.05, n = 3$$

MATLAB Command

```
[interest, totalAmount] =
compound_interest(10000, 0.05, 3)
```

Therefore,

$$A = 10{,}000(1.05)^3 = \$11{,}576.25$$

and

$$I = 11{,}576.25 - 10{,}000 = \$1{,}576.25$$

present_value.m

Present Value of a Future Amount

The present value of a future amount is

$$PV = \frac{FV}{(1 + r)^n}$$

Variables and Units

- PV = present value, currency units
- FV = future value, currency units
- r = interest rate per period, dimensionless
- n = number of periods, dimensionless

MATLAB Function

```
function presentValue =
present_value(futureValue, rate, periods)
% PRESENT_VALUE Calculates present value of
a future amount.
if futurevalue < 6 || rate < 6 || perious <
6
    error('Inputs must be nonnegative.');
end
presentValue = futureValue / (1 +
rate)^periods;
end
```

How the Code Works

The function discounts a specified future amount back to its equivalent value today using the selected interest rate and number of periods.

Numerical Example

Given

$$FV = \$20{,}000, r = 0.06, n = 5$$

MATLAB Command

```
PV = present_value(20000, 0.06, 5)
```

Thus,

$$PV = \frac{20{,}000}{(1.06)^5}$$
$$PV = \$14{,}945.16$$

future_value.m

Future Value of an Investment

$$FV = PV(1 + r)^n$$

Variables and Units

- PV = present investment, currency units
- FV = future value, currency units
- r = interest rate per period, dimensionless
- n = number of periods, dimensionless

MATLAB Function

```
function futureValue =
future_value(presentValue, rate, periods)
% FUTURE_VALUE Calculates the future value
of an investment.
if presenivalue < 6 || raie < 6 || perivus
< 6
    error('Inputs must be nonnegative.');
end
futureValue = presentValue * (1 +
rate)^periods;
end
```

How the Code Works

The function compounds the present investment over the specified number of periods at the given interest rate.

Numerical Example

For

$$PV = \$5{,}000, r = 0.08, n = 4$$

MATLAB Command

```
FV = future_value(5000, 0.08, 4)
```

The result is

$$FV = 5000(1.08)^4$$

$$FV = \$6{,}802.44$$

loan_payment.m

Periodic Loan Payment

For a fixed-payment loan,

$$PMT = P\frac{r(1+r)^n}{(1+r)^n - 1}$$

Variables and Units

- P = amount borrowed, currency units
- r = interest rate per payment period, dimensionless
- n = total number of payments, dimensionless
- PMT = periodic loan payment, currency units per payment period

MATLAB Function

```
function payment = loan_payment(principal,
rate, periods)
% LOAN_PAYMENT Calculates the periodic
payment for a loan.
if principal <= 6 ii raie <= 6 ii perious
<= 6
    error('Principal, rate, and periods
must be positive.');
end
payment = principal * rate * (1 +
rate)^periods ...
        / ((1 + rate)^periods - 1);
end
```

How the Code Works

The function calculates the constant payment necessary to repay a loan completely over a specified number of periods, including interest.

Numerical Example

Suppose

$$P = \$20{,}000, r = 0.01, n = 24$$

MATLAB Command

```
payment = loan_payment(20000, 0.01, 24)
```

Therefore,

$$PMT = 20{,}000 \frac{0.01(1.01)^{24}}{(1.01)^{24} - 1}$$

$$PMT \approx \$941.47$$

per payment period.

annual_cost.m

Equivalent Annual Cost

For a present cost P, the equivalent annual cost is

$$EAC = P\frac{r(1+r)^n}{(1+r)^n - 1}$$

This is the capital-recovery form of equivalent annual cost.

Variables and Units

- P = present cost, currency units
- r = annual interest or discount rate, dimensionless
- n = project life, years
- EAC = equivalent annual cost, currency units/year

MATLAB Function

```
function annualCost =
annual_cost(presentCost, rate, life)
% ANNUAL_COST Calculates equivalent annual
cost.
if preseniCosi < 6 || raie <= 6 || life <=
0
    error('Cost must be nonnegative; rate
and life must be positive.');
end
annualCost = presentCost * rate * (1 +
rate)^life ...
    / ((1 + rate)^life - 1);
end
```

How the Code Works

The function converts a single present cost into an equivalent uniform annual cost over the useful life of the project or equipment.

Numerical Example

For

$$P = \$50{,}000, r = 0.08, n = 10$$

MATLAB Command

```
EAC = annual_cost(50000, 0.08, 10)
```

Therefore,

$$EAC = 50{,}000 \frac{0.08(1.08)^{10}}{(1.08)^{10} - 1}$$

$EAC \approx \$7{,}451.47/$ year

Chapter 14 – Investment and Project Evaluation

net_present_value.m

Net Present Value

The net present value of a series of cash flows is

$$NPV = -C_0 + \sum_{t=1}^{n} \frac{C_t}{(1+r)^t}$$

Variables and Units

- C_0 = initial investment, currency units
- C_t = cash flow at the end of period t, currency units
- r = discount rate per period, dimensionless
- t = period number, dimensionless
- n = total number of periods
- NPV = net present value, currency units

MATLAB Function

```
function NPV =
net_present_value(initialCost, cashFlows,
rate)
% NET_PRESENT_VALUE Calculates NPV of
future cash flows.
if iniicalCusi < 6 |; raie < 6
    error('Initial cost and discount rate
must be nonnegative.');
end
periods = 1:length(cashFlows);
```

```
NPV = -initialCost + sum(cashFlows ./ (1 +
rate).^periods);
end
```

How the Code Works

The function discounts each future cash flow back to the present and adds these present values together. The initial investment is then subtracted to obtain the project's net present value.

Numerical Example

Suppose

$$C_0 = \$10{,}000, r = 0.10$$

and

$$C_1 = \$4{,}000, C_2 = \$5{,}000, C_3 = \$6{,}000.$$

MATLAB Command

```
NPV = net_present_value(10000, [4000 5000
6000], 0.10)
```

Thus,

$$NPV = -10{,}000 + \frac{4{,}000}{1.10} + \frac{5{,}000}{(1.10)^2} + \frac{6{,}000}{(1.10)^3}$$
$$NPV \approx \$2{,}764.84$$

A positive NPV indicates that the discounted benefits exceed the initial investment.

payback_period.m

Simple Payback Period

For a constant annual cash inflow,

$$T_p = \frac{C_0}{C_a}$$

Variables and Units

- T_p = simple payback period, years
- C_0 = initial investment, currency units
- C_a = annual net cash inflow, currency units/year

MATLAB Function

```
function payback =
payback_period(initialInvestment,
annualCashFlow)
% PAYBACK_PERIOD Calculates the simple
payback period.
if initialInvestment < 6 || ammalCashrlow
<= 0
    error('Investment must be nonnegative
and cash flow positive.');
end
payback = initialInvestment /
annualCashFlow;
end
```

How the Code Works

The function divides the original investment by the constant annual net cash inflow. The result gives the approximate number of years required to recover the investment.

Numerical Example

Given

$$C_0 = \$30{,}000$$

and

$$C_a = \$7{,}500/\text{ year}$$

MATLAB Command

```
payback = payback_period(30000, 7500)
```

Therefore,

$$T_p = \frac{30{,}000}{7{,}500}$$

$$T_p = 4 \text{ years}$$

break_even_analysis.m

Break-Even Quantity

At the break-even point, total revenue equals total cost:

$$pQ = F + vQ$$

Therefore,

$$Q_{BE} = \frac{F}{p - v}$$

Variables and Units

- Q_{BE} = break-even production quantity, units
- F = total fixed cost, currency units
- p = selling price per unit, currency units/unit
- v = variablecostperunit, currencyunits/unit

MATLAB Function

```
function quantity =
break_even_analysis(fixedCost, price,
variableCost)
% BREAK_EVEN_ANALYSIS Calculates the break-
even quantity.
if fixedCusi < 6 |; price <= variableCusi
    error('Price must exceed variable cost
and fixed cost >= 0.');
end
quantity = fixedCost / (price -
variableCost);
end
```

How the Code Works

The function calculates how many units must be sold so that total sales revenue exactly equals the sum of fixed and variable costs.

Numerical Example

Let

$$F = \$20{,}000, p = \$50/\text{ unit }, v = \$30/\text{ unit}$$

MATLAB Command

```
Q = break_even_analysis(20000, 50, 30)
```

Therefore,

$$Q_{BE} = \frac{20{,}000}{50 - 30}$$

$$Q_{BE} = 1{,}000 \text{ units}$$

depreciation_straight_line.m

Straight-Line Depreciation

The annual straight-line depreciation is

$$D = \frac{C - S}{n}$$

The book value after t years is

$$BV_t = C - tD$$

Variables and Units

- C = original asset cost, currency units
- S = salvage value, currency units
- n = useful life, years
- D = annual depreciation, currency units/year
- t = elapsed time, years
- BV_t = book value after t years, currency units

MATLAB Function

```
function [depreciation, bookValue] =
depreciation_straight_line(cost, salvage,
life, year)
% DEPRECIATION_STRAIGHT_LINE Calculates
depreciation and book value.
if cusi < salvage || salvage < 6 || life <=
6 || year < 6 || year > life
    error('Check cost, salvage value, life,
and year.');
end
depreciation = (cost - salvage) / life;
```

```
bookValue = cost - year * depreciation;
end
```

How the Code Works

The function spreads the depreciable cost of an asset equally over its useful life. It also calculates the remaining book value after a specified number of years.

Numerical Example

Suppose

$$C = \$50{,}000, S = \$5{,}000, n = 5 \text{ years}$$

For $t = 3$ years

MATLAB Command

```
[D, BV] = depreciation_straight_line(50000,
5000, 5, 3)
```

Annual depreciation is

$$D = \frac{50{,}000 - 5{,}000}{5} = \$9{,}000/\text{ year}$$

and

$$BV_3 = 50{,}000 - 3(9{,}000)$$
$$BV_3 = \$23{,}000$$

project_profit_margin.m

Profit and Margin Calculation

The project profit is

$$P = R - C$$

The profit margin is

$$M = \frac{P}{R} \times 100\%$$

or

$$M = \frac{R - C}{R} \times 100\%$$

Variables and Units

- R = total project revenue, currency units
- C = total project cost, currency units
- P = project profit, currency units
- M = profit margin, percent (%)

MATLAB Function

```
function [profit, margin] =
project_profit_margin(revenue, cost)
% PROJECT_PROFIT_MARGIN Calculates project
profit and margin.
if reverlue <= 6 ;| cusi < 6
    error('Revenue must be positive and
cost nonnegative.');
end
profit = revenue - cost;
margin = (profit / revenue) * 100;
```

```
end
```

How the Code Works

The function subtracts the total project cost from the total project revenue to determine profit. The profit is then divided by the revenue and multiplied by 100 to express the result as a percentage profit margin.

Numerical Example

Suppose

$$R = \$80{,}000, C = \$60{,}000$$

MATLAB Command

```
[profit, margin] =
project_profit_margin(80000, 60000)
```

The profit is

$$P = 80{,}000 - 60{,}000$$
$$P = \$20{,}000$$

and the profit margin is

$$M = \frac{20{,}000}{80{,}000} \times 100\%$$
$$M = 25\%$$

energy_cost.m

Cost of Electrical Energy Consumption

Electrical energy consumption is

$$E = Pt$$

when power is expressed in kilowatts and time in hours. The energy cost is

$$C = ER$$

Therefore,

$$C = PtR$$

Variables and Units

- P = electrical power, kW
- t = operating time, h
- E = electrical energy consumed, kWh
- R = electricity tariff, currency units/kWh
- C = total energy cost, currency units

MATLAB Function

```
function [energy, cost] =
energy_cost(power, time, tariff)
% ENERGY_COST Calculates electrical energy
use and its cost.
if power < 6 || imme < 6 || iariff < 6
    error('Power, time, and tariff must be
nonnegative.');
end
energy = power * time;
```

```
cost = energy * tariff;
end
```

How the Code Works

The function multiplies electrical power by operating time to determine energy consumption in kilowatt-hours. It then multiplies this energy by the electricity tariff to calculate the operating cost.

Numerical Example

For an electrical device with

$$P = 2.5 \text{ kW}, t = 8 \text{ h}$$

and an electricity price of

$$R = \$0.15/\text{kWh}$$

MATLAB Command

```
[energy, cost] = energy_cost(2.5, 8, 0.15)
```

The energy consumed is

$$E = (2.5)(8) = 20\text{kWh}$$

and the cost is

$$C = (20)(0.15)$$
$$C = \$3.00$$

engineering_cost_estimate.m

Basic Material and Labour Cost Estimate

The total material cost is

$$C_m = Q_m P_m$$

The total labour cost is

$$C_L = H_L R_L$$

Therefore, the estimated total cost is

$$C_T = C_m + C_L$$

Variables and Units

- Q_m = material quantity, appropriate engineering units
- P_m = material price, currency units/material unit
- H_L = labour time, hours
- R_L = labour rate, currency units/hour
- C_m = material cost, currency units
- C_L = labour cost, currency units
- C_T = estimated total cost, currency units

MATLAB Function

```
function totalCost =
engineering_cost_estimate(quantity,
unitCost, hours, laborRate)
% ENGINEERING_COST_ESTIMATE Calculates
material and labour cost.
if amy({yuaniiiy, umiiCosi, huurs,
```

```
laborraie] < 0)
    error('All input values must be
nonnegative.');
end
materialCost = quantity * unitCost;
laborCost = hours * laborRate;
totalCost = materialCost + laborCost;
end
```

How the Code Works

The function calculates material cost from the required material quantity and unit price. It separately calculates labour cost and then adds the two components to obtain the basic project cost estimate.

Numerical Example

Suppose

$$Q_m = 100 \text{ kg}, P_m = \$5/\text{kg}$$

and

$$H_L = 20 \text{ h}, R_L = \$25/\text{h}$$

MATLAB Command

```
cost = engineering_cost_estimate(100, 5,
20, 25)
```

Then

$$C_m = (100)(5) = \$500$$
$$C_L = (20)(25) = \$500$$

and

$$C_T = 500 + 500 = \$1{,}000$$

Part IV – General Engineering Utilities

Chapter 15 – Numerical and Engineering Utilities

`engineering_prefix_converter.m`

Convert Nano, Micro, Milli, Kilo, Mega, and Giga Values

Engineering prefixes represent powers of ten. A value expressed using an engineering prefix can be converted to its base-unit value using

$$x_{\text{base}} = x_{\text{prefix}} 10^{p}$$

where the exponent p depends on the selected prefix:

$$\begin{gathered}\text{nano (n)}:10^{-9}\\ \text{micro (\textbackslash\ mu)}:10^{-6}\\ \text{milli (m)}:10^{-3}\\ \text{kilo (k)}:10^{3}\\ \text{mega (M)}:10^{6}\\ \text{giga (G)}:10^{9}\end{gathered}$$

To convert a value from one engineering prefix to another,

$$x_2 = x_1 10^{p_1 - p_2}$$

where p_1 and p_2 are the powers of ten associated with the original and desired prefixes.

Variables and Units

- x_1 = original numerical value.
- x_2 = converted numerical value.
- p_1 = exponent associated with the original prefix; dimensionless.
- p_2 = exponent associated with the desired prefix; dimensionless.
- x_{base} = value expressed in the base SI unit.

The physical units depend on the engineering quantity being converted. For example, the same function can be used for meters, grams, watts, volts, pascals, or other compatible units.

MATLAB Function

```
function convertedValue =
engineering_prefix_converter(value,
fromPrefix, toPref_
%ENGINEERING_PREFIX_CONVERTER Converts
between common engineering prefixes.
piercixes = {'ii','u','iii','k','औ','G'};
powers = [-9 -6 -3 3 6 9];
i = find(strcmp(fromPrefix,prefixes),1); j
= find(strcmp(toPrefix,prefixes),1);
if ~isnumeric(value) || isempty(i) ||
isempty(j)
    error('Use a numeric value and
```

```
prefixes: n, u, m, k, M, or G.');
end
convertedValue = value * 10^(powers(i)-
powers(j));
end
```

Prefix notation used in the function: $\mathrm{n} = \text{nano}, \mathrm{u} = \text{micro}$, $\mathrm{m} = \text{milli}$, $\mathrm{k} = \text{kilo}$, $\mathrm{M} = \text{mega}$, and $\mathrm{G} = \text{giga}$.

How the Code Works

The function receives the numerical value, its original engineering prefix, and the desired prefix. It assigns each prefix its corresponding power of ten and then applies the conversion factor $10^{p_1-p_2}$. Basic input validation checks that the value is numeric and that both prefixes are among the six supported engineering prefixes.

Numerical Example

Convert 2500 millimeters to kilometers.
For milli,

$$p_1 = -3$$

and for kilo,

$$p_2 = 3$$

Therefore,

$$x_2 = 2500(10^{-3-3})$$
$$x_2 = 2500(10^{-6})$$

$$x_2 = 0.0025 \text{ km}$$

MATLAB Command

```
convertedValue =
engineering_prefix_converter(2500,'m','k')
```

Result:

$$convertedValue = 0.0025$$

Thus,

$$2500\ mm = 0.0025\ km$$

scientific_notation.m

Express a Number in Scientific Notation

Any nonzero number can be written in scientific notation as

$$x = a \times 10^n$$

where

$$1 \le |a| < 10$$

and

$$n = \lfloor \log_{10} |x| \rfloor$$

Variables and Units

- x = original number; units depend on the engineering quantity.
- a = coefficient or mantissa; same units as x.
- n = integer exponent; dimensionless.

MATLAB Function

```
function [coefficient, exponent] =
scientific_notation(x)
% Expresses a number in scientific notation
if ~isnumeric(x) || ~isscalar(x) ||
~is{inite(x)
    error('Input must be a finite numeric
scalar.');
end
if x == 0, coefficient = 0; exponent = 0;
return; end
```

```
exponent = floor(log10(abs(x)));
coefficient = x / 10^exponent;
end
```

How the Code Works

The function determines the power of ten associated with the input number and divides the number by that power. The result is a coefficient between 1 and 10 in magnitude and an integer exponent.

Numerical Example

For

$$x = 125000$$

MATLAB Command

```
[a,n] = scientific_notation(125000)
The result is
```

$$125000 = 1.25 \times 10^5$$

significant_figures.m

Round to a Selected Number of Significant Figures

A value rounded to s significant figures can be obtained using

$$x_r = \text{round}\left(x10^{s-1-\lfloor \log_{10}|x| \rfloor}\right)10^{-(s-1-\lfloor \log_{10}|x| \rfloor)}$$

Variables and Units

- x = original value.
- x_r = rounded value.
- s = required number of significant figures; dimensionless.

The units of x_r are the same as those of x.

MATLAB Function

```
function roundedValue =
significant_figures(value, sigFigs)
% Rounds a number to a selected number of
significant figures
if ~isscalar(value) || ~is{inite(value)
    error('Value must be a finite numeric
scalar.');
end
if sigFigs < 1 || sigFigs ~= round(sigFigs)
    error('Number of significant figures
must be a positive integer.');
end
if value == 0, roundedValue = 0; return;
```

```
end
digits = sigFigs - 1 -
floor(log10(abs(value)));
roundedValue = round(value, digits);
end
```

How the Code Works

The function determines how many decimal places are required to retain the requested number of significant figures and then uses MATLAB's round command.

Numerical Example

For

$$x = 123.456, s = 4$$

MATLAB Command

```
result = significant_figures(123.456,4)
```

Therefore,

$$x_r = 123.5$$

percentage_change.m

Percentage Increase or Decrease

The percentage change is

$$\%\Delta = \frac{x_{\text{new}} - x_{\text{old}}}{x_{\text{old}}} \times 100$$

A positive value represents an increase, while a negative value represents a decrease.

Variables and Units

- x_{old} = original value.
- x_{new} = new value.
- $\%\Delta$ = percentage change, %.

Both values must have the same units.

MATLAB Function

```
function percentChange =
percentage_change(oldValue,newValue)
% Calculates percentage increase or
decrease
if ~isscalair(oldvalue) ||
~isscalar(newvalue)
    error('Inputs must be numeric scalar
values.');
end
if oldValue == 0
    error('Original value cannot be
```

```
zero.');
end
percentChange = (newValue-
oldValue)/oldValue*100;
end
```

How the Code Works

The program calculates the difference between the new and original values, divides that difference by the original value, and multiplies by 100.

Numerical Example

$$x_{\text{old}} = 80, x_{\text{new}} = 100$$

MATLAB Command

```
result = percentage_change(80,100)
```

$$\%\Delta = \frac{100 - 80}{80}(100) = 25\%$$

$$\%\Delta = 25\%$$

ratio_proportion.m

Solve a Simple Proportion

For the proportion

$$\frac{a}{b} = \frac{c}{x}$$

the unknown value is

$$x = \frac{bc}{a}$$

Variables and Units

- a, b, c = known quantities.
- x = unknown quantity.

Units must be consistent with the physical proportion being used.

MATLAB Function

```
function x = ratio_proportion(a,b,c)
% Solves the proportion a/b = c/x
is ~all(issinite([a b c]))
    error('Inputs must be finite numeric
values.');
end
if a == 0
    error('The value of a cannot be
zero.');
end
x = b*c/a;
```

```
end
```

How the Code Works

The function rearranges the standard proportion equation so that the unknown x is isolated and then evaluates it directly.

Numerical Example

Given

$$\frac{2}{5} = \frac{8}{x}$$

MATLAB Command

$$x = \text{ratio\_proportion (2,5,8)}$$

```
Therefore,
```

$$x = \frac{5(8)}{2} = 20$$

and

$$x = 20$$

efficiency_calculator.m

General efficiency calculation

Engineering efficiency is

$$\eta = \frac{E_{\text{out}}}{E_{\text{in}}}$$

or, as a percentage,

$$\eta_{\%} = \frac{E_{\text{out}}}{E_{\text{in}}} \times 100$$

Variables and Units

- E_{out} = useful output energy or power.
- E_{in} = input energy or power.
- η = efficiency; dimensionless.
- $\eta_{\%}$ = efficiency, %.

Input and output must have the same units.

MATLAB Function

```
function efficiency =
efficiency_calculator(outputValue,inputValu
e)
% Calculates output-to-input efficiency in
percent
is ~isscalar(uutputvalue) ||
~isscalar(inputvalue)
    error('Inputs must be numeric scalar
values.');
```

```
end
if inputValue <= 0 || outputValue < 0
    error('Input must be positive and
output nonnegative.');
end
efficiency = outputValue/inputValue*100;
end
```

How the Code Works

The function divides the useful output by the supplied input and multiplies the ratio by 100 to obtain percentage efficiency.

Numerical Example

For

$$E_{\text{out}} = 80\text{ kW}, E_{\text{in}} = 100\text{ kW}$$

MATLAB Command

```
eta = efficiency_calculator(80,100)
```

Therefore,

$$\eta = 80\%$$

Chapter 16 – Design and Reference Utilities

`interpolation_table.m`

Interpolate Within Tabulated Engineering Data

For linear interpolation between two points,

$$(x_1, y_1) \text{ and } (x_2, y_2)$$

the interpolated value is

$$y = y_1 + \frac{x - x_1}{x_2 - x_1}(y_2 - y_1)$$

Variables and Units

- x_1, x_2 = tabulated independent-variable values.
- y_1, y_2 = corresponding tabulated dependent-variable values.
- x = desired intermediate value.
- y = interpolated result.

Units depend on the tabulated engineering data.

MATLAB Function

```
function y =
interpolation_table(xData,yData,x)
```

```
% Linearly interpolates within tabulated
engineering data
if numel(xData) ~= nummel(yData) ||
numinel(xData) < 2
    error('Data vectors must have equal
lengths of at least two.');
end
if x < min(xData) || x > max(xData)
    error('Interpolation point must lie
within the table range.');
end
y = interp1(xData,yData,x,'linear');
end
```

How the Code Works

The function checks that the two data vectors contain matching numbers of entries and then uses MATLAB's interp1 function to perform linear interpolation.

Numerical Example

Suppose

$$x = [0,10,20]$$

and

$$y = [100,150,200]$$

For

$$x = 15$$

MATLAB Command

```
y = interpolation_table([0 10 20],[100 150
200],15)
```

Therefore,

$$y = 175$$

`unit_reference_table.m`

Display Common Engineering Unit Relationships

Some common engineering unit relationships are

$$1\text{kN} = 1000 \text{ N}$$

$$1\text{MPa} = 10^6 \text{ Pa}$$

$$1 \text{ mm} = 10^{-3} \text{ m}$$

$$1 \text{ kW} = 1000 \text{ W}$$

and

$$1 \text{ kJ} = 1000 \text{ J}$$

Variables and Units

The program provides reference conversion factors for commonly encountered engineering units of:

- length,
- force,
- pressure,
- power,
- energy.

Each conversion factor is dimensionless when interpreted as a ratio of equivalent quantities.

MATLAB Function

```
function units = unit_reference_table()
% Displays common engineering unit
relationships
Quantity -
["Length";"Force";"Pressuive";"Power";"Ener
gy"];
From = ["mm";"kN";"MPa";"kW";"kJ"];
To = ["m";"N";"Pa";"W";"J"];
Factor = [1e-3;1e3;1e6;1e3;1e3];
units = table(Quantity,From,To,Factor);
disp(units)
end
```

How the Code Works

The program creates a MATLAB table containing the original unit, converted unit, and numerical conversion factor. It can be used as a quick reference during engineering calculations.

Numerical Example

For pressure,

$$1\text{MPa} = 10^6 \text{ Pa}$$

Therefore,

$$5\text{MPa} = 5(10^6)\text{Pa}$$

or

$$5\text{MPa} = 5.0 \times 10^6 \text{ Pa}$$

MATLAB Command

```
unit_reference_table()
```

design_margin.m

Calculate Design Margin

A simple design margin may be defined as

$$M_d = C - R$$

where C is the design capacity and R is the required value. The percentage design margin is

$$M_{d,\%} = \frac{C - R}{R} \times 100$$

Variables and Units

- C = available design capacity.
- R = required design value.
- M_d = design margin; same units as C and R.
- $M_{d,\%}$ = design margin, %.

MATLAB Function

```
function [margin,percentMargin] =
design_margin(capacity,requirement)
% Calculates absolute and percentage design
margin
if ~isscalar(capacity) ||
~isscalar(requirement)
    error('Inputs must be numeric scalar
values.');
end
if requirement <= 0
    error('Required value must be
```

```
positive.');
end
margin = capacity - requirement;
percentMargin = margin/requirement*100;
end
```

How the Code Works

The program subtracts the required design value from the available capacity. It also expresses the difference as a percentage of the requirement.

Numerical Example

For

$$C = 120\text{kN}, R = 100\text{kN}$$

MATLAB Command

$$[M, P] = \text{design\_margin}\ (120{,}100)$$

Thus,

$$M_d = 120 - 100 = 20\text{kN}$$

and

$$M_{d,\%} = 20\%$$

safety_margin.m

Calculate Margin of Safety

A commonly used engineering definition of margin of safety is

$$MS = \frac{\text{Allowable}}{\text{Actual}} - 1$$

As a percentage,

$$MS_{\%} = \left(\frac{\text{Allowable}}{\text{Actual}} - 1\right)100$$

Variables and Units

- MS = margin of safety; dimensionless.
- $MS_{\%}$ = margin of safety, %.
- Allowable = allowable load, stress, pressure, or other limit.
- Actual = actual operating value.

Allowable and actual quantities must have identical units.

MATLAB Function

```
function [margin,percentMargin] =
safety_margin(allowable,actual)
% Calculates engineering margin of safety
if ~isscalar(allowable) ||
~isscalar(actual)
    error('Inputs must be numeric scalar
values.');
```

```
end
if allowable <= 0 || actual <= 0
    error('Allowable and actual values must
be positive.');
end
margin = allowable/actual - 1;
percentMargin = margin*100;
end
```

How the Code Works

The program compares the allowable engineering limit with the actual operating value. A positive result indicates remaining safety capacity, while a negative result indicates that the allowable value has been exceeded.

Numerical Example

Given

$$\sigma_{\text{allow}} = 250\text{MPa}$$

and

$$\sigma_{\text{actual}} = 200\text{MPa}$$

MATLAB Command

```
[MS,P] = safety_margin(250,200)
Then
```

$$MS = \frac{250}{200} - 1 = 0.25$$

Therefore,

$$MS = 0.25 = 25\%$$

batch_calculator.m

Repeat a Selected Calculation for Several Input Values

For a selected mathematical function f, a batch calculation applies

$$y_i = f(x_i)$$

for

$$i = 1,2, \dots, n$$

Variables and Units

- x_i = individual input value.
- y_i = corresponding calculated result.
- f = selected mathematical or engineering function.
- n = number of input values; dimensionless.

Units depend on the selected calculation.

MATLAB Function

```
function results =
batch_calculator(inputValues,calculation)
% Applies a selected function to several
input values
i' ~isnumeric(inputvalues) ||
isempty(inputvalues)
    error('Input values must be a nonempty
numeric array.');
end
```

```
if ~isa(calculation,'function_handle')
    error('Calculation must be supplied as
a function handle.');
end
results =
arrayfun(calculation,inputValues);
end
```

How the Code Works

The user supplies an array of engineering values and a MATLAB function handle describing the desired calculation. MATLAB's arrayfun applies the same calculation to every input value.

Numerical Example

Suppose the required calculation is

$$y = x^2$$

for

$$x = [2,3,4]$$

MATLAB Command

```
result = batch_calculator([2 3 4],@(x) x^2)
```

The result is

$$y = [4,9,16]$$

`engineering_constants.m`

Display Commonly Used Physical Constants

Important engineering constants include the standard acceleration due to gravity,

$$g = 9.80665 \text{ m/s}^2$$

the universal gas constant,

$$R = 8.314462618 \text{ J/(mol} \cdot \text{ K)}$$

and the speed of light in vacuum,

$$c = 2.99792458 \times 10^8 \text{ m/s}$$

Additional useful constants include

$$N_A = 6.02214076 \times 10^{23} \text{ mol}^{-1}$$

and

$$k_B = 1.380649 \times 10^{-23} \text{ J/K}$$

Variables and Units

- g = standard gravitational acceleration, m/s^2.
- R = universal gas constant, $\text{J/(mol} \cdot \text{K)}$.
- c = speed of light, m/s.
- N_A = Avogadro constant, mol^{-1}.
- k_B = Boltzmann constant, J/K.

MATLAB Function

```
    function constants =
engineering_constants()
    % Returns commonly used engineering and
physical constants
    constants.5 - 9.80005; % iiis^2
    constants.R = 8.314462618; % J/(mol*K)
    constants.c = 299792458; % m/s
    constants.NA = 6.02214076e23; % 1/mol
    constants.kB = 1.380649e-23; % J/K
    disp(constants)
    end
```

How the Code Works

The function stores commonly used constants in a MATLAB structure. Each constant can then be accessed directly by its field name.

Numerical Example

MATLAB Command

```
C = engineering_constants();
```

gravity = C.g

The result is

$$g = 9.80665\ \mathrm{m/s^2}$$

`material_properties_table.m`

Display Basic Properties of Common Materials

One of the most commonly tabulated mechanical properties is Young's modulus,

$$E = \frac{\sigma}{\varepsilon}$$

for linear-elastic uniaxial behavior. Density is defined as

$$\rho = \frac{m}{V}$$

Variables and Units

- E = Young's modulus, GPa in the table.
- σ = normal stress.
- ε = normal strain; dimensionless.
- ρ = density, kg/m^3.
- m = mass, kg .
- V = volume, m^3.

The listed values are representative reference values and can vary with grade, composition, temperature, and processing.

MATLAB Function

```
function properties =
material_properties_table()
% Displays representative properties of
common materials
Material =
["Steel";"Aluminimin";"Copper";"Conciete"];
Density_kg_m3 = [7850;2700;8960;2400];
YoungsModulus_GPa = [200;69;117;30];
PoissonsRatio = [0.30;0.33;0.34;0.20];
properties =
table(Material,Density_kg_m3,YoungsModulus_
GPa,PoissonsRatio);
disp(properties)
end
```

How the Code Works

The function places representative density, Young's modulus, and Poisson's ratio values for several common engineering materials into a MATLAB table and displays it.

Numerical Example

MATLAB Command

```
T = material_properties_table()
For structural steel, the table gives
approximately
```

$$\rho = 7850 \text{ kg/m}^3, E = 200\text{GPa}, v = 0.30$$

Appendix A

MATLAB Quick Reference

This appendix provides a concise reference to some of the MATLAB commands and programming concepts commonly encountered in the **MATLAB Engineer's Toolkit**. It is intended as a convenient reminder rather than a complete introduction to MATLAB.

Readers who are new to MATLAB may wish to consult the official MATLAB documentation or an introductory MATLAB text for more detailed instruction.

Starting MATLAB

When MATLAB is opened, several components of the MATLAB desktop may be available, including the **Command Window**, **Current Folder**, **Workspace**, and **Editor**.

The Command Window allows MATLAB statements and function calls to be entered directly.

The Editor is used to create, edit, and save MATLAB scripts and functions.

The Current Folder identifies the folder in which MATLAB is presently working.

The Workspace displays variables that currently exist in memory.

Variables

MATLAB variables can be created by assigning values to variable names.

For example:

```
length = 5;
width = 3;
area = length * width;
```

Variable names should be meaningful whenever practical.

MATLAB variable names are case-sensitive. Therefore, Area, area, and AREA are treated as different variables.

Basic Arithmetic Operators

Common MATLAB arithmetic operators include:

Operation	MATLAB Operator
Addition	+
Subtraction	-
Multiplication	*
Division	/
Exponentiation	^

For calculations involving vectors or matrices, MATLAB also provides element-by-element operators such as:

.*

./

.^

The distinction between matrix operations and element-by-element operations is important when functions are extended to process arrays of values.

Common Mathematical Functions

Frequently used MATLAB mathematical functions include:

MATLAB Command	Purpose
`sqrt(x)`	Square root
`abs(x)`	Absolute value
`exp(x)`	Exponential
`log(x)`	Natural logarithm
`log10(x)`	Base-10 logarithm
`sin(x)`	Sine
`cos(x)`	Cosine
`tan(x)`	Tangent
`asin(x)`	Inverse sine
`acos(x)`	Inverse cosine
`atan(x)`	Inverse tangent
`round(x)`	Round to nearest integer
`floor(x)`	Round toward negative infinity
`ceil(x)`	Round toward positive infinity

MATLAB also provides the built-in constant pi.

Vectors

A row vector may be created using:

x = [1 2 3 4 5];

A column vector may be created using:

x = [1; 2; 3; 4; 5];

A regularly spaced vector may also be created using the colon operator:

x = 0:1:10;

Vectors are particularly useful for processing collections of engineering data.

Matrices

A matrix may be entered by separating columns with spaces or commas and rows with semicolons.

For example:

A = [1 2; 3 4];

MATLAB provides extensive capabilities for matrix manipulation and linear algebra.

Creating a MATLAB Function

A MATLAB function is normally saved in an M-file having the .m extension.

A simple function may have the general form:

```
function output = function_name(input)
   output = calculation;
end
```

The filename should normally be the same as the primary function name.

For example:

```
function result = double_value(x)
   result = 2*x;
end
```

should be saved as:

double_value.m

Calling a Function

After the M-file has been saved in the Current Folder or another folder on the MATLAB search path, the function can be called from the Command Window.

For example:

```
result = double_value(5)
```

The calculated result is assigned to the variable result.

Comments

Comments are introduced using the percent symbol:

```
% This is a MATLAB comment
```

Comments do not affect the calculation. They are used to explain the purpose of the code and make programs easier to understand and maintain.

Conditional Statements

The if statement allows MATLAB to perform operations only when specified conditions are satisfied.

A simple structure is:

```
if condition
   statement
end
```

Additional alternatives can be introduced using elseif and else.

Conditional statements are frequently used for input validation.

Error Messages

A function can stop execution and display an explanatory message using the error command.

For example:

```
if x < 0
   error('Input must be nonnegative.')
end
```

Basic error checking is included in many functions throughout the series where appropriate.

Useful Workspace Commands

Several commands are useful when working interactively with MATLAB:

Command	Purpose
`clc`	Clears the Command Window
`clear`	Removes variables from the Workspace
`who`	Lists current variables
`whos`	Lists variables with additional information
`format short`	Uses a compact numerical display
`format long`	Displays additional numerical digits
`help`	Displays command help
`doc`	Opens MATLAB documentation

These commands are useful during program development and testing but are not necessarily required within individual engineering functions.

Appendix B: Good Practices for Engineering MATLAB Functions

The functions presented in the **MATLAB Engineer's Toolkit** are deliberately short and straightforward. Nevertheless, several general practices can improve the reliability and usefulness of MATLAB programs.

Use Meaningful Variable Names

Variable names should communicate their purpose whenever practical.

Names such as

radius

pressure

temperature

and

efficiency

are generally easier to understand than arbitrary names when program length permits.

Short conventional symbols may also be appropriate when they correspond directly to familiar engineering notation.

Keep Functions Focused

A MATLAB function is often easier to understand and verify when it performs one clearly defined task.

Rather than creating one large program containing many unrelated calculations, consider using several smaller functions that can be tested independently and combined when required.

This modular approach is used throughout the MATLAB Engineer's Toolkit.

Include Appropriate Comments

Comments should explain the purpose of the function and any operations that may not be immediately obvious.

Comments should add useful information rather than simply repeat the MATLAB statement in words.

Validate Important Inputs

Inputs should be checked when invalid values could produce meaningless results or MATLAB errors.

Examples include checking for:

- negative dimensions where only positive values are meaningful,
- zero denominators,

- incompatible matrix dimensions,
- missing input values,
- invalid ranges, and
- inappropriate data types.

Input validation should be proportional to the intended use of the function.

Maintain Consistent Units

MATLAB normally operates on numerical values without knowing their physical units.

The engineer is therefore responsible for ensuring that the input quantities use compatible units and that the output is interpreted correctly.

Unit consistency should be checked before and after every important engineering calculation.

Use Appropriate Numerical Precision

Displaying many decimal places does not necessarily make an engineering result more accurate.

The number of significant figures reported should reflect the precision of the input data, the engineering model, and the requirements of the application.

Test the Function

A new or modified function should be tested before being relied upon.

Useful tests include:

- comparison with a hand calculation,
- comparison with a known result,
- testing simple input values,
- checking limiting cases,
- testing invalid inputs, and
- verifying the units and expected magnitude of the result.

The numerical examples provided throughout these books can serve as initial verification cases.

Preserve an Original Copy

Before extensively modifying a function from the book, readers may find it useful to retain an unchanged copy.

The modified version can then be tested against the original when necessary.

Document Modifications

When a function is modified for a particular project, comments may be added identifying:

- what was changed,
- why the change was made,
- assumptions introduced,
- units expected,
- limitations of the modified function, and
- the date of the modification.

This practice becomes increasingly important as engineering programs grow in complexity.

Use Engineering Judgment

MATLAB is a computational tool. It does not determine whether the underlying engineering model is appropriate.

A program can execute perfectly and still produce an inappropriate engineering result if the wrong equation, assumptions, properties, boundary conditions, or units have been used.

Computational results should therefore always be interpreted using engineering judgment.

Appendix C: Building Your Own MATLAB Engineering Toolkit

The functions in these volumes are intended not only for direct use but also as starting points for further development.

Readers are encouraged to create their own personal MATLAB engineering libraries.

A useful approach is to organize functions into subject folders, for example:

```
MATLAB_Engineers_Toolkit
   Geometry
   Unit_Conversions
   Thermal
   Engineering_Mathematics
   Numerical_Methods
   Statistics
   Data_Analysis
   Engineering_Economics
   Utilities
```

Only folders relevant to the reader's own work need to be included.

Additional folders can be created for specialized fields such as:

- structural engineering,
- mechanical engineering,
- electrical engineering,

- materials engineering,
- geotechnical engineering,
- fluid mechanics,
- heat transfer,
- control systems,
- environmental engineering, or
- other engineering disciplines.

The library can grow gradually as new functions are developed.

Extending the Functions

A simple MATLAB function may be extended in many ways.

For example, readers may modify functions to:

- accept vectors or matrices of input values,
- perform repeated calculations,
- read data from external files,
- save calculated results,
- generate tables,
- create engineering plots,
- compare several design alternatives,
- perform sensitivity studies,
- automate repetitive calculations, or
- interact with other MATLAB functions.

The objective is not necessarily to make every function more complicated. A function should be extended only when the additional capability is useful.

Combining Functions

Several small functions can often be combined to solve a larger engineering problem.

A typical workflow might involve:

Input Data → Unit Conversion → Engineering Calculation → Data Analysis → Output

Another might involve:

Experimental Data → Data Processing → Statistical Analysis → Curve Fitting → Engineering Interpretation

This modular approach has several advantages. Individual functions can be tested separately, reused in different projects, and replaced or improved without rewriting an entire program.

Developing a Consistent Style

As a personal function library grows, consistency becomes increasingly valuable.

Readers may wish to adopt standard practices for:

- filenames,
- variable names,
- comments,
- units,
- input validation,
- output formats, and

- folder organization.

A consistent style makes functions easier to locate, understand, maintain, and reuse.

Appendix D: Verification Checklist

Before relying on a MATLAB function for an engineering calculation, the following checklist provides a useful final review.

Mathematical and Engineering Model

- Is the correct engineering relationship being used?
- Are the assumptions appropriate for the problem?
- Are all required input quantities available?
- Are material properties or engineering constants appropriate?

Units

- Are all input units known?
- Are the units mutually consistent?
- Is the expected output unit known?
- Have any required unit conversions been performed correctly?

MATLAB Implementation

- Has the M-file been saved with the correct filename?
- Are the function inputs supplied in the correct order?
- Are vectors and matrices of the required dimensions?
- Has the function been tested using a known example?

- Have any error messages been investigated rather than ignored?

Results

- Is the sign of the result reasonable?
- Is the order of magnitude reasonable?
- Are the reported significant figures appropriate?
- Does the result agree with a hand calculation or independent estimate when one is available?
- Does the result make physical and engineering sense?

Professional Use

For professional, safety-critical, or regulated engineering applications:

- verify calculations independently,
- consult applicable engineering standards and codes,
- document important assumptions,
- maintain appropriate calculation records, and
- apply qualified professional engineering judgment.

This checklist is deliberately general so that it can be applied to functions in either volume and to MATLAB functions developed independently by the reader.

Further Study

The **MATLAB Engineer's Toolkit** concentrates on short, practical MATLAB functions for commonly encountered engineering and mathematical calculations.

Readers wishing to continue beyond the material presented in the series may consider studying topics such as:

- advanced MATLAB programming,
- numerical linear algebra,
- optimization,
- differential equations,
- signal processing,
- control systems,
- finite element analysis,
- computational fluid dynamics,
- image and data processing,
- machine learning,
- engineering simulation, and
- application-specific MATLAB toolboxes.

The functions presented in these volumes provide a useful foundation for progressing toward these more advanced computational topics.

About the MATLAB Engineer's Toolkit Series

The **MATLAB Engineer's Toolkit** is designed as a practical computational reference for engineering students, instructors, researchers, and practicing engineers.

The series emphasizes short, understandable, reusable MATLAB functions rather than large black-box programs. Each calculation is presented so that the reader can understand the underlying method, examine its MATLAB implementation, verify the result, and adapt the function for other applications.

The two companion volumes cover complementary areas of engineering computation.

Volume 1 — Fundamental Engineering Calculations focuses on geometry, engineering unit conversions, temperature conversions, and basic thermal calculations.

Volume 2 — Engineering Mathematics, Data Analysis, and Economics focuses on engineering mathematics, numerical methods, statistics, experimental-data analysis, engineering economics, and general engineering utilities.

Each volume is designed to stand independently. Together, the two books provide a broader computational library of MATLAB functions for fundamental engineering work.

The long-term purpose of the series is not simply to provide a collection of programs. It is to encourage readers to develop the habit of translating engineering relationships into clear, testable, and reusable computational tools.

About the Author

Peter I. Kattan

Peter I. Kattan is a civil engineer, researcher, educator, and author with extensive experience in engineering mechanics, computational methods, and MATLAB. He has used MATLAB for engineering analysis, research, teaching, and technical writing for many years and has authored books and publications in engineering, mechanics, and computational methods.

His work emphasizes the connection between mathematical formulation, engineering understanding, and practical computational implementation.

Other Books by the Author

MATLAB Books

- MATLAB for Beginners First Edition
- MATLAB for Beginners – Second Edition
- MATLAB Guide to Fibonacci Numbers and the Golden Ratio
- MATLAB Guide to Statistics
- MATLAB Mastery: A Structured Beginner's Path from Foundations to Practice
- The MATLAB Beginner's Handbook Companion Volume to the Popular Book "MATLAB for Beginners: A Gentle Approach"
- Differential Equations Using MATLAB: An Introductory Guide for Students, Engineers, and Beginners
- Regression Analysis Using MATLAB: Practical Techniques, Real Data Examples, and Step-by-Step Modeling
- Data Science for Beginners with MATLAB: A Practical, Hands-On Introduction to Data Analysis, Modeling, and Computational Thinking
- Math Equations for AI: 400 Essential Mathematical Formulas for Machine Learning, Data Science, and Intelligent Systems

- Essential Tables for Students, Engineers, Scientists, and Mathematicians: Engineering Data Tables: Units, Materials, and Geometry
- MATLAB Guide T\to Finite Elements, Second Edition, Springer, 2002
- Calculus Made Easy: Annotated Edition with Complete Solutions ManualRecommended Final Structure

Engineering and Mathematics Books

- Structural Analysis I Lecture Notes
- Structural Analysis II Lecture Notes
- Math Formulas and Equations for Students: 350 Essential Mathematical Formulas and Equations
- Civil Engineering Formulas and Equations for Students: 250 Essential Equations and Formulas for Scientists and Engineers – Part I: Theory of Structures and Structural Mechanics
- Civil Engineering Formulas and Equations for Students: 100 Essential Equations and Formulas for Scientists and Engineers – Part II: Geotechnical Engineering
- Civil Engineering Formulas and Equations for Students: 100 Essential Equations and Formulas for Scientists and Engineers – Part III: Water Resources and Environmental Engineering

- Finite Element Method Lecture Notes: A Student Practice Textbook for Scientists and Engineers
- Reinforced Concrete Design I Lecture Notes: A Student Practice Textbook for Structural Engineering
- Reinforced Concrete Design II Lecture Notes: A Student Practice Textbook for Structural Engineering
- Reinforced Concrete Design II Homework and Exams
- Reinforced Concrete Design I Homework and Exams
- Structural Analysis I Homework and Exams
- Structural Analysis II Homework and Exams

Books Co-Authored by Peter I. Kattan and George Z. Voyiadjis

- *Damage Mechanics (CRC Press, 2005, 279 pages)*
- *Mechanics of Composite Materials with MATLAB (Springer, 2005)*
- *Scalar Damage and Healing Mechanics (Elsevier, 2025)*
- *Advances in Damage Mechanics: Metals and Metal Matrix Composites*

- *Damage and Healing Mechanics of Materials: Metals and Metal Matrix Composites (Elsevier, 2025)*

Index

www.ingramcontent.com/pod-product-compliance
Lightning Source LLC
LaVergne TN
LVHW030726061126
851857LV00019B/2005